SAGE was founded in 1965 by Sara Miller McCune to support the dissemination of usable knowledge by publishing innovative and high-quality research and teaching content. Today, we publish more than 750 journals, including those of more than 300 learned societies, more than 800 new books per year, and a growing range of library products including archives, data, case studies, reports, conference highlights, and video. SAGE remains majority-owned by our founder, and on her passing will become owned by a charitable trust that secures our continued independence.

Los Angeles | London | Washington DC | New Delhi | Singapore

The Sardar Sarovar Project

Thank you for choosing a SAGE product! If you have any comment, observation or feedback, I would like to personally hear from you. Please write to me at contactceo@sagepub.in

—Vivek Mehra, Managing Director and CEO,
SAGE Publications India Pvt Ltd, New Delhi

Bulk Sales

SAGE India offers special discounts for purchase of books in bulk. We also make available special imprints and excerpts from our books on demand.

For orders and enquiries, write to us at

Marketing Department
SAGE Publications India Pvt Ltd
B1/I-1, Mohan Cooperative Industrial Area
Mathura Road, Post Bag 7
New Delhi 110044, India
E-mail us at marketing@sagepub.in

Get to know more about SAGE, be invited to SAGE events, get on our mailing list. Write today to marketing@sagepub.in

This book is also available as an e-book.

The Sardar Sarovar Project

ASSESSING ECONOMIC AND SOCIAL IMPACTS

S. Jagadeesan
and
M. Dinesh Kumar

⑤SAGE www.sagepublications.com
Los Angeles • London • New Delhi • Singapore • Washington DC

First published in 2015 by

SAGE Publications India Pvt Ltd
B1/I-1 Mohan Cooperative Industrial Area
Mathura Road, New Delhi 110 044, India
www.sagepub.in

SAGE Publications Inc
2455 Teller Road
Thousand Oaks, California 91320, USA

SAGE Publications Ltd
1 Oliver's Yard, 55 City Road
London EC1Y 1SP, United Kingdom

SAGE Publications Asia-Pacific Pte Ltd
3 Church Street
#10-04 Samsung Hub
Singapore 049483

Published by Vivek Mehra for SAGE Publications India Pvt Ltd, typeset in 10/13 Times New Roman by RECTO Graphics, Delhi and printed at Sai Print-o-pack, New Delhi.

Library of Congress Cataloging-in-Publication Data Available

ISBN: 978-93-515-0126-8 (HB)

The SAGE Team: N. Unni Nair, Sandhya Gola, Rajib Chatterjee, and Vinitha Nair

Contents

List of Tables   vii
List of Figures   xi
Foreword by Cecilia Tortajada   xiii
Acknowledgments   xvii

1. Introduction   1
2. Setting the Global Context   16
3. Social Benefits and Impacts: An Analysis of the
   Sardar Sarovar Project   39
4. Narmada River Basin and Sardar Sarovar Project   60
5. Sardar Sarovar Project and Improving Groundwater
   Regime in Overexploited Regions of Gujarat   74
6. Socioeconomic Impact of Canal Irrigation   95
7. Drinking Water Supplies from Narmada:
   Socioeconomic Impacts   159
8. Indirect Impacts of Irrigation and Drinking Water Supply 183
9. Environmental Externalities of the Sardar Sarovar Project 221
10. The Socioeconomic Impacts on Displaced Population   235
11. Maximizing Future Benefits and Minimizing Negative
    Impacts from the Sardar Sarovar Project   257
12. Conclusion   274

Bibliography   296
Index   309
About the Authors   313

List of Tables

4.1   Expected and Current Utilization of Irrigation and Power
Potential and Water Resources of the Narmada Basin   63

4.2   District-wise Irrigation Benefits from the SSP (Planned)   70

4.3   Progress in the Construction of Canal Network of the SSP
(as on 2012)   72

5.1   Change in the SWL (January)   78

5.2   Change in the SWL (May)   79

5.3   Change in the SWL (October)   79

5.4   TDS (January)   83

5.5   TDS (May)   84

5.6   TDS (October)   84

6.1   Gujarat's Share of the Volumetric Water Allocation from
the Narmada River Basin and Utilization over the Years   97

6.2   Changes in Sources of Irrigation and Area Irrigated
with the Introduction of Narmada Waters in Canal
Command Areas   101

6.3   Changes in the Source of Irrigation and Irrigated Area
with Irrigation through Canal Lift   102

6.4   Percentage Area under Crops in Different Seasons before
and after Narmada Waters   104

6.5   Percentage Area under Crops in Different Seasons before
and after Irrigation through Canal Lifting   106

6.6   Change in Crop Yield (kg per ha) under Canal Irrigation   109

6.7   Change in Yield (kg per ha) of Crops with the Introduction
of Water from Narmada Canal through Lifting   112

6.8   The Cost of Cultivation of Canal Irrigators (₹ per ha)   119

6.9   Change in the Cost of Cultivation (₹ per ha) of Crops
Irrigated through Canal Lifting   121

6.10   Average Farm Gate Price of Different Crops (₹ per kg)
       in the Canal Command Areas   125
6.11   Changes in Income from Various Crops in Canal
       Command Area after the Introduction of Narmada Waters   126
6.12   Average Farm Gate Prices of Different Crops (₹ per kg)
       in Areas Irrigated by Canal Lift   128
6.13   Change in the Average Net Income of Canal Lift
       Irrigators after the Introduction of Narmada Waters   129
6.14   Average Net Returns from Crop Production in
       Non-command Area   136
6.15   Change in Size of Animal Holding in Narmada Canal
       Command Areas per Farm Household   140
6.16   Change in the Size of Animal Holding in Areas
       Irrigated by Narmada Canal Lift   140
6.17   Annual Income (₹) of Dairy Farmers in the Canal
       Command Areas   141
6.18   Annual Income (₹) of Dairy Farmers Using Canal
       Water Using Lift   142
6.19   Overall Farm Surplus of Canal Irrigators   143
6.20   Overall Farm Surplus of Farmers Irrigating through
       Canal Lift   145
6.21   Changes in Wage Employment in Agriculture Due to
       Canal Irrigation (Days)   148
6.22   Impact of Canal Irrigation on Household Consumption
       Expenditure (₹ per Year)   150
6.23   Impact of Irrigation on Household Consumption
       Expenditure (₹ per Annum) of Canal Lift Irrigators   152

7.1   Progress in Drinking Water Supply Coverage in Gujarat
       Villages over Time   161
7.2   Changing Source of Water Supply with the
       Introduction of Water from Narmada Pipeline:
       Rural Jamnagar and Jamnagar Town   163
7.3   Changes in Water Supply Source with the Introduction
       of Narmada Water in Rural Bhuj and Bhuj Town   165
7.4   Frequency of Water Supply before and after
       Narmada Canal-based Piped Water Supply   166

7.5   Total Distance Traveled per Day for Collection
      of Water (m)   167
7.6   Impact of Water Supply on the Average Time Spent on
      Collecting Water for Domestic Uses in Different Locations 169
7.7   Seasonal Domestic Water Consumption by the Sample
      Households: Rainy Season   172
7.8   Seasonal Domestic Water Consumption by the Sample
      Households: Winter   173
7.9   Seasonal Water Consumption by the Sample
      Households: Summer   174
7.10 Water Quality Impacts of Narmada Canal-based Water
      Supply on Different Household Uses   175
7.11 Changes in Health Expenditure Due to Water-related
      Diseases   178

8.1   Average Groundwater Use for Crop Production in the
      Well Command Area (m^3 per ha)   187
8.2   Changes in the Incidence of Well Failures and Well
      Command Area in Narmada Command after the
      Introduction of Canal Irrigation   188
8.3   Change in Area under Various Crops (ha) in Well
      Commands after Narmada   190
8.4   Percentage Change in Area under Various Crops of
      Well Irrigators after Narmada Waters   192
8.5   Change in the Yield of Crops Grown in the Well
      Commands in Six Locations, Due to Narmada Waters   193
8.6   Change in the Cost of Cultivation of Crops Grown in the
      Well Commands in Six Locations, Due to Narmada Waters   195
8.7   Change in the Net Income of Well-owning Farmers from
      Crop Production   196
8.8   Percentage Change in the Net Income of Well Irrigators
      from Crops after Narmada   198
8.9   Average Size of Animal Holding of Well Owners before
      and after Narmada Canal   200
8.10 Average Income of Well-owning Farmers from Dairy
      Production After Narmada Canal (₹ per Animal per
      Lactation) before and after Narmada   201

8.11 Changes in Farm Surplus of Well Irrigators in Canal
Command Areas 203

8.12 Changes in Wage Rate (₹ per Day) of Farm Laborers in
the Five Locations over Time across Seasons for Male and
Female Agricultural Laborers 205

8.13 Indirect Benefit of Canal Irrigation through Improved
Water Supplies 210

8.14 Positive Externality of Improved Water Supply 213

8.15 The Cost of Electricity for Water Supply 215

9.1 Characteristics of Induced River Flow and the Perceived
Benefits 230

9.2 Travel Cost Calculations 231

10.1 Norms for R&R as per the NWDT Award and Norms
Followed by the States 239

10.2 Total Number of PAFs in Different States in the Upper
Catchment of Reservoir 241

10.3 Resettlement Sites of Sardar Sarovar Dam Oustees in
Gujarat 242

10.4 Status of Rehabilitation of Sardar Sarovar Dam Oustees 243

10.5 Distribution of the Productive Assets amongst the Oustees
from the Three States 244

10.6 Changes in Crop Choices, Cropping Pattern, and Irrigation
Pattern with Resettlement 247

10.7 Change in the Yield of Cereal Crops 248

10.8 Change in Animal Holding of Households with
Resettlement 248

10.9 Change in Degree of Access to Formal Water Supply
Sources with Resettlement 250

10.10 Change in Access to Medicare with Resettlement 251

10.11 Changes in Modes of Food Collection 253

10.12 Change in Overall Level of Satisfaction with Life with
Resettlement 254

11.1 Water Productivity of Various Crops 271

12.1 Direct and Indirect Benefits from the Sardar Sarovar
Narmada Project 276

List of Figures

3.1 A Flowchart Depicting the Direct and Indirect Impacts
of the SSP   41

3.2 Map of Gujarat with Arrows Showing Study Locations   54

4.1 Per Capita Water Availability (m³ per Annum) in
Different Regions of Gujarat   66

5.1 Average Rise/Fall in SWL (January)   82
5.2 Average Rise/Fall in SWL (May)   82
5.3 Average Rise/Fall in SWL (October)   82
5.4 Average Rise/Fall in TDS Value (January)   87
5.5 Average Rise/Fall in TDS Value (May)   87
5.6 Average Rise/Fall in TDS Value (October)   88

6.1 Growth in Agricultural GSDP in Gujarat, Constant
Prices (1999–2000)   99
6.2 Source-wise Change in Irrigated Area (ha)   103
6.3 Impact of Narmada on the Yield of Crops (kg per ha)
Irrigated by Canal (Kharif)   110
6.4 Impact of Narmada on Yield of Crops (kg per ha)
Irrigated through Canal Lift (Kharif)   116
6.5 Number of Rain-fed and Irrigated Crops before and
after Waters Narmada (Canal Command Area)   117
6.6 Number of Rain-fed and Irrigated Crops before and
after Narmada Canal Lift   118
6.7 Current Net Income (₹ per ha) for Canal Irrigators
(Monsoon)   132

6.8 Current Net Income (₹ per ha) from Various Crops
for Canal Lift Irrigators (Monsoon)   133

6.9 Current Net Income (₹ per ha) for Canal Irrigator
(Winter and Summer)   134

6.10 Current Net Income (₹ per ha) from Various Crops
for Canal Lift Irrigators (Winter and Summer)   135

6.11 Increase in Wage Employment (No. of Days) in Different
Seasons, across Locations in Narmada Command   149

6.12 Impact of Canal Irrigation on Family Expenditure   151

7.1 Per Capita Daily Domestic Water Use (Liters)   171

8.1 Fluctuation in the Water Level in Wells (ft),
Pre- and Post-Narmada   186

8.2 Average Well Command Area (ha) in Narmada Canal
Command before and after Narmada Waters   189

8.3 The Intensity of Groundwater Use (m³ per ha) in
Well Commands   194

8.4 Estimated Effective Increase in Wage Rates in
Real Terms (₹)   206

8.5 Quality of Water Used for Drinking and Cooking   211

8.6 Quality of Water for Domestic Purpose   211

9.1 Power Generation from the SSP 2005–2011 (Million Units)   223

9.2 Average Pesticide Consumption per Hectare of
Net Sown Area in Gujarat   228

12.1 Per Capita Annual Reservoir Storage (m³) in
Selected Countries   292

12.2 Sustainable Water Use Index versus per Capita
Reservoir Storage   292

Foreword

Originally conceived by the late Sardar Vallabh Bhai Patel in 1946–1947, the Sardar Sarovar Project (SSP) on the Narmada River has been one of the most controversial projects in the recent history of India. This colossal multipurpose project covers four states in India: Gujarat, Madhya Pradesh (MP), Maharashtra, and Rajasthan, and is intended to provide water for human, agricultural, industrial, and hydropower uses at a massive scale. Once finalized, the project is expected to meet the drinking water needs of 131 urban centers and 9,633 villages as far as Saurashtra, Kutch, and North Gujarat, irrigate 1.8 million hectares of land, and generate approximately 1,450 MW annually.

The SSP has faced a myriad of economic, financial, social, and environmental-related challenges from its inception. The size of the project and the extent of its potential social, economic, and environmental impacts are not only positive but also negative, which have resulted in strong opposition by various social and environment activist groups.

The demands of the groups have included, and rightly so, the improved social and environmental performance of the states and national government and also of the World Bank as the lending institution. On the one hand, the pressure of some of the non-governmental organizations (NGOs) for better social and environmental policies has been positive since it resulted in the improvement of the rehabilitation package of the "oustees," and on thousands of hectares of areas being reforested. At the same time, however, there have been NGOs whose objective seemed to have been politically motivated and whose contribution to the improvement in the quality of life of the local population did not necessarily translate in visible benefits. One of the main objectives of their movement was to attract international media attention which was positive for them and negative for the project, and they got both of them.

Reports in the media regarding opposition to the project were such that in October 1989 the One Hundred First Congress of the United States held a congressional hearing before the Subcommittee on Natural Resources, Agricultural Research, and Environment. The objective of this hearing was to consider "the serious social and environmental issues which surround the construction of the Sardar Sarovar dam, part of the Narmada Valley Development Project in India." In addition to the overall impacts of the project, the World Bank decision-making and integrity, its social and environmental guidelines, and the extent to which these were implemented by the countries whose projects were financed by the Bank were also discussed at length during the hearing. From India, to talk to the US Congress were Professor Vijay Paranjpye, the economist who had carried out a cost–benefit analysis of the project for the Indian National Trust for Art and Cultural Heritage, Ms Medha Patkar, the sociologist representing the Narmada Bachao Andolan (NBA), and Mr Girish Patel, the President of Lok Adhikar Sangh, the People's Right Organization. During the hearing, the benefits of the project were put in serious doubt, its negative impacts were extensively discussed, and the overall performance of the World Bank was critically questioned, giving a general impression of a project that was negative from all points of view.[1]

In 1991, the World Bank commissioned an independent review of the environmental and resettlement aspects of the project. The Morse Report, as it was subsequently known, was very critical of the project. It essentially recommended that the World Bank suspended funding until resettlement, rehabilitation, and environmental considerations were properly addressed. The World Bank subsequently dismissed some of the review's concerns but agreed with other ones and finally requested the Indian government to comply with certain social and environmental standards in a record time of six months. The end result was that the Government of India, after consultation with the governments of the four

---

[1] Hearing before the Subcommittee on Natural Resources, Agricultural Research, and Environment of the Committee on Science, Space, and Technology, the US House of Representatives, One Hundred First Congress: First Session, No. 68, October 24, 1989, printed by the US Government Printing Office, Washington, 1990.

states involved, announced that it would withdraw the loan application. The construction of the project continued with national funds.

SSP became the lighting road for opposition by social and environment activist groups with major consequences as the ones mentioned before. There is no doubt that there are compelling reasons to maximize the benefits and minimize the negative impacts of large water projects irrespective of the pressures from environmental and social activists. Nonetheless, the discussions in the international arena have often gone out of bounds. Many international activist groups demand that no new dams are constructed, irrespective of their overall total benefit to the society. While some of these groups are genuinely interested on the populations affected, many others follow only their own dogmatic views. As acknowledged by policy-makers, practitioners, and also activists all over the world, the concern for the people and the environment, both genuine and faked, has made that many times basic needs and the rights of the low-income majority of the population of the developing world are ignored.

The validity of the arguments for and against dams by governmental institutions and NGOs cannot be resolved one way or the other because of the lack of past and present post-project evaluations of the economic, social, and environmental impacts (both positive and negative) of large dams in different parts of the world. Therefore, only anecdotal information can be used to justify or refute the arguments made by the proponents or the opponents of the large dams. Until and unless objective and reliable impact assessments and post-project evaluations are carried out, and then analyzed and disseminated, no definite conclusions on the overall benefits and costs of dams can be drawn.

The importance of water development projects for the socioeconomic development of developing countries cannot be denied. Nevertheless, anecdotal evidence supports the fact that both the positive and adverse social and environmental impacts of large water projects have been seriously underestimated. It is the absence of objective post-project evaluations what has resulted in the lack of conclusive and definitive statements on the overall impacts of large dams on the society and the environment.

In this regard, the book *The Sardar Sarovar Project: Assessing Economic and Social Impacts* is an example of much needed post-evaluation assessments of large projects. Properly implemented, such a large project provides a unique opportunity to improve the quality of life of millions of people who many times live in disadvantageous situations. The overall impacts of such a colossal project are very significant both in reach and extent. With so many interrelated impacts, when reading the book, one certainly agrees with the authors on the importance of carefully understanding the impacts in order to be able to evaluate them properly. This book contributes to the understanding of the benefits of the project and also provides rich sources of information for further studies on social, economic, and environmental topics in the area of influence of the project.

As to whether the positive impacts of the project will be sustainable as mentioned by the authors, this will be a matter of time, effort by all parties involved and careful monitoring of the progress made. Regular ex-post analysis of large projects, such as the present one, represents one of the best instruments to identify and maximize positive impacts, minimize negative ones, and recognize emerging concerns so that appropriate policies can be formulated and timely actions can be taken in order to target them as necessary.

As mentioned by former South African president, Late Nelson Mandela, during the launch of the World Commission on Dams Report,

[t]he problem, though, is not the dams. It is the hunger. It is the thirst. It is the darkness of a township. It is townships and rural huts without running water, lights or sanitation. It is the time wasted gathering water by hand. There is a real pressing need for power in every sense of the word. Rather than single out dams for excessive blame, or credit, we must learn to answer: "It is all of us!" All of us must wrestle with the difficult questions we face.

Dr Cecilia Tortajada
President, Third World Centre for Water Management, Mexico

Acknowledgments

At the outset, we would like to thank Sardar Sarovar Narmada Nigam Limited (SSNNL) for supporting an important study *Realistic vs Mechanistic: Assessing the Economic and Social Benefits from SSP During the Year 2010*, which formed the basis for several chapters contained in this book. Particularly, we would like to express our deep gratitude to Dr J. N. Singh, IAS, the Managing Director of SSNNL, and a known irrigation expert. He took special interest in getting the final manuscript ready for publication.

The Madras Institute of Development Studies (MIDS), Chennai, participated in the study in reviewing the methodology and fine-tuning the report with useful comments. Particularly, we thank Dr L. Venkatachalam, Associate Professor, MIDS, and Professor Maria Saleth, Director, MIDS, for their valuable inputs. The study report was presented in a workshop organized jointly by SSNNL and the Institute for Resource Analysis and Policy, the organization which undertook the study. The workshop was organized at the Centre for Environmental Planning and Technology (CEPT) University, Ahmedabad, on August 7, 2012. It was attended by several academics, administrators, and policy-makers from the state and central agencies connected with water and development in India. Dr Mihir Shah, Planning Commission Member, National Advisory Council, in addition to several senior officials of SSNNL participated in the workshop. The suggestions and comments received during the workshop were later on incorporated in the report, including the suggestion to include a chapter on the socioeconomic impact of the project on dam oustees and also to compare the farming enterprise of those who receive irrigation from the project with those who are outside the project's designated command area.

The draft manuscript was subsequently reviewed by three experts in the field of water resources, one from SSNNL and two from outside. They were Dr Mukesh B. Joshi, Chief Engineer, SSNNL; Professor R. Maria Saleth, Director, MIDS; and Shri Avinash Tyagi, Secretary General, ICID, New Delhi. The authors are extremely thankful to them for their valuable suggestions and comments. Dr Cecilia Tortajada, President, Third World Centre for Water Management, who wrote the "Foreword" for the book, also offered several valuable comments and suggestions for sharpening the focus of the manuscript. The comments from all these scholars were incorporated for enriching the document.

Several officers from SSNNL played an important role in making this project possible. To mention a few are Shri K. Srinivas, IAS, Joint Managing Director (Finance); Dr K. D. Acharya, Deputy Engineer; Shri R. G. Acharya, Officer on Special Duty and Deputy General Manager, Public Relations. We thank them profusely. The authors would also like to place on record their thanks to the contributions of researchers from Institute for Resource Analysis and Policy, Shri Nitin Bassi, Shri Niranjan Vedantam, Shri Shiv Ram Kishan. and Dr M. V. K. Sivamohan, and its consultants, Shri Kairav Trivedi and Dr O. P. Singh, without whose help the project would not have come to a fruition.

S. Jagadeesan
M. Dinesh Kumar

1

Introduction

Large irrigation projects are targets of increased criticism worldwide for the negative social and environmental effect they are likely to cause. Critiques argue that the costs outweigh their intended benefits (Biswas and Tortajada, 2001; Shah and Kumar, 2008; Verghese, 2001). The SSP constructed across Narmada River is no exception to this and, as a result, remains one of the most controversial water development projects in independent India (Verghese, 2001; Shah and Kumar, 2008). However, the available analyses of performance of SSP were based on narrow objectives and with ideological overtones (see for instance Parasuraman *et al.*, 2010; Shah *et al.*, 2009; Talati and Shah, 2004). But, a recent work, which examined the agricultural growth in Gujarat, displayed the enormous significance of the SSP in stabilizing agricultural production and reviving the growth. This study as well as the earlier one by Shah and Kumar (2008) emphasized the need for analyzing the social, economic, and environmental benefits from the project using broader set of criteria.

The underlying concern is that as several decades rolled by since the project was initially conceptualized and planned, the social, economic, and environmental context had undergone a metamorphosis. While waterlogging was projected as a negative externality of the project, with secular decline in groundwater levels throughout the arid and semiarid Gujarat, this has become a nonissue (Shah and Kumar, 2008). On the contrary, seepage and return flows from irrigation are likely to contribute to the improvement of groundwater balance (Ranade and Kumar, 2004;

Shah and Kumar, 2008), sustaining well yields and reducing the energy requirements for lifting water, and, therefore, becoming now a positive externality. With increased recharge of water from rivers and canals and irrigated fields, a near permanent solution to the problems of both the exposure to poor quality groundwater (with high fluoride and total dissolved solids [TDS]) for drinking and the failure of drinking water wells in rural areas is found in central and north Gujarat.

Further, myriad of villages in Gujarat started experiencing problems of acute shortage of drinking water due to groundwater overexploitation and quality deterioration with fluoride and salinity, and their number had increased remarkably over the years (Kumar, 2007). Many thousands of villages have fully run out of local freshwater sources during summer, unable to meet even the basic survival needs, causing large-scale rural out-migration. As Gleick (2000) pointed out, right to water for basic survival is even more basic and vital than some of the more explicit human rights already acknowledged by the international community. If that is the case, the impact of Narmada waters, when supplied for rural and urban drinking so far, is likely to be far greater than expected as it promotes not only social advancements but also human rights. Significant reduction in children's absenteeism from schools, increase in wages earned by women in villages, and reduction in the expenditure on health are also visible.

On the energy front, the crisis is far more critical today. With rapid industrialization underway in Gujarat and with an increase in the projected energy demand for the future, the opportunity cost of having to live with power deficits is much larger. Needless to say, the actual economic cost of fossil fuel-based power generation is much higher, with growing concerns of climate change and global warming. In this context, the SSP helps Gujarat and MP produce clean and cheap energy.

With declining per capita cereal production (NRAA, 2011) and their prices skyrocketing (Chowdhury, 2011), the country as a whole is faced with a major food crisis, and food insecurity (Kumar *et al.*, 2012) is posing a new challenge to the very social fabric. One of the reasons for these phenomena is the growing irrigation water scarcity and mounting energy costs, which are driving the farmers out from traditional food crops to more remunerative cash crops, creating, thereby, local food shortages. This is particularly true for north Gujarat, where groundwater, which used to be the major source of irrigation, is heavily overexploited. If one goes by the initial reports, the introduction of water from SSP had resulted in bumper production of wheat, bananas, potatoes, and paddy in the state during the past 4–5 years, which is also blessed with four consecutive good monsoons.

All this shows that apart from its life-sustaining value (Joshi and Kapadia, 2010), the potential economic contribution of SSP in terms of meeting sustainable development goals such as improving water security and health, food and energy security, and economic growth, while reducing the "environmental stresses" caused by economic activities, is substantial (Desai and Joshi, 2008; Vyas, 2001; Shah and Kumar, 2008). Hence, the criteria for evaluating the project benefits need a thorough relook. On the other hand, with groundwater buildup in the command area, conjunctive use of both surface water and groundwater in the command becomes achievable. But, the critics continue to evaluate the performance in relation only to the intended benefits such as irrigation by canals, power generation, and water supply for drinking and industry.

The impact of SSP vis-à-vis the social and economic benefits they generate need to be evaluated carefully. Also, the social and environmental implications of the project are to be fully understood. The import of Narmada water into the stressed rivers of north Gujarat had huge positive ecological/environmental effects. Water from Narmada released through canals is not only used for irrigation, but also for several socioeconomic production functions such as domestic use, washing, bathing, livestock drinking, and fisheries. Impacts brought out by such uses are extremely important for projects like SSP, which involve large-scale interbasin transfer of water from water-abundant to water-stressed regions.

Water Security and Economic Growth: National and Global Context

The impacts of large water resource projects in the developing countries need to be examined from the point of view of water security, energy security, and economic growth needs (Perry, 2001a). India, whose economy is in transition, is an illustrative example. Large parts of India are naturally water-scarce, but are agriculturally prosperous regions which produce surplus food grains for export to other regions. These regions face shortage of water for irrigation, especially during summer months, with the limited availability of surface water and groundwater depletion (Amarasinghe *et al.*, 2004). Water insecurity problems become severe during droughts. Whereas the water-abundant eastern India is not only agriculturally backward, but also does not offer much potential for future growth due to acute shortage of arable land which can be brought under irrigated cultivation. Future growth in agriculture can come from only naturally water-scarce regions, and this would be possible only if water security is ensured (Kumar *et al.*, 2012). This is also true for regions such as Gujarat, which have large areas under semiarid and arid conditions and where agriculture is heavily dependent on availability of irrigation water, and agricultural output and growth are hit by frequent droughts. With rapid industrialization, the manufacturing output is heavily dependent on water, which is a key input.

India's energy demand is also growing rapidly as manifested by the exponential increase in electricity consumption per capita witnessed during 1970–1971 to 2010–2011. The compounded annual growth rate in aggregate electricity use was 7.2 percent during this period. This comes from all four energy use sectors, that is, irrigated agriculture, domestic sector in both urban and rural areas, and manufacturing and service sector, though the growth in demand has been the highest (Compound Annual Growth Rate 15.7 percent) in the industrial sector. But, the growth in installed capacity of the electricity utilities grew only at the rate of 6.6 percent, from 16,271 MW to 206,526 MW. But, nearly, 78 percent of the additional 789 billion units of annual electricity generation achieved by 2010–2011 came from thermal power, which is nonrenewable, whereas the contribution of hydropower, which forms clean energy source, was only 19.6 percent (source: based on data in NSO, 2012). This raises serious concerns about long-term energy security, and also environment.

Globally, there is a growing debate on the linkages between water, growth, and development. According to some scholars, increased investment in water projects such as irrigation, hydropower, water supply, and sanitation acts as engines of growth in the economy while supporting progress in human development (for instance, see Biswas *et al.*, 2004; Briscoe, 2005; HDR, 2006). They harp on the need for investment in water infrastructure and institutions. Grey and Sadoff (2007) suggest that there is a minimum platform of water security, achieved through the right combination of investment in water infrastructure and institutions and governance, which is essential if poor countries are to use water resources effectively and efficiently to achieve rapid economic growth to benefit vast numbers of their population. They suggest an S-curve for growth impacts of investment in water infrastructure and institutions in which returns continue to be nil for early investments. They argue that for poor countries, which experience highly variable climates, the level of investment required to reach the tipping point of water security[1] would be much higher as compared to countries which fall in temperate climate with low variability. But, they suggest that for developing countries, the returns on investment in infrastructure would be higher than in management, and vice versa in the case of developed countries. Kumar (2009) empirically showed that improving water security drives economic growth through the human development route. Further, in arid and semiarid tropical countries, increasing per capita water storage is the key to improving water security, which in turn can drive economic growth.

Many environmental groups, on the other hand, advocate small water projects managed by the communities. The solutions advocated are: watershed management, small water harvesting interventions (Iyer, 2012), community-based water supply systems, and micro-hydroelectric projects (Dharmadhikary, 2005; D'Souza, 2002).

The proponents of sustainable development paradigms believe that the ability of a country to sustain its economic growth depends on the extent to which natural resources, including water, are put to efficient use through technologies and institutions, thereby reducing the stresses on environmental resources (Drexhage and Murphy, 2010). Here, the focus is on initiating institutional and policy reforms in the water sector. An alternative view suggests that countries would be able to tackle their water scarcity and other problems relating to water environment at advanced stages of economic development (Shah and Koppen, 2006).

---

[1] Beyond which the investment in water infrastructure and institutions yields positive growth impacts.

They argue that standard approaches to water management in terms of policies and institutions work when water economies become formal, which are found at an advanced stage of economic development of nations.

International literature provides many clues to the fact that water security has the potential to promote inclusive growth. For instance, access to safe water and sanitation can partly determine income. The marginal productivity per unit of water measured in terms of good health, longevity, or income is much greater for the poor than for the rich (World Water Council, 2000; Jha, 2010; van Koppen *et al.*, 2009). Secured water for irrigation, while enabling the poor landholding communities to grow food for their own consumption (Kumar, 2003), would generate sufficient employment in rural areas and lower cereal prices (Perry, 2001a) while supplying cereals to the markets. In poor and developing countries, where large section of the population depends on agriculture and rural wage labor, this is likely to have significant distributional effects on income. Thus, improved water situation leads to better human health and environmental sanitation, food security and nutrition, livelihoods, and, thereby, greater access to education for the poor (UNDP, 2006).

Analyzing the Direct and Indirect Benefits of Large Water Projects

The current research is about the direct and indirect impacts of large water resource projects on the socioeconomic and environmental landscapes of the regions they serve. This was analyzed through a comprehensive study of one of the most magnificent yet controversial, multipurpose water projects in the developing world, which is the SSP on Narmada River in India, which is still under implementation. The analysis is set in the context of growing debate on the role of large dams in the social and economic development of countries, whose countryside is largely dependent on agriculture as a major source of livelihood.

The overall objective of the research was to undertake a comprehensive evaluation of the current and likely future impacts of the Sardar Sarovar Narmada Project in Gujarat on the social, economic, and ecological/environmental fronts, including the potential for conjunctive use of surface water and groundwater in the SSP command. The specific objectives were as follows: evaluate the direct economic and socioeconomic impacts of Narmada water supplies through canals in command areas and outside; assess the socioeconomic impacts of the introduction of Narmada waters for drinking water supplies in rural and urban areas of Gujarat; evaluate the physical and environmental impacts of introducing irrigation water in the command areas; and carry out an analysis of the economic benefits of the SSP, which, along with the direct economic costs and benefits, includes the positive and negative externalities induced by the use of Narmada water for irrigation and drinking and hydropower generation from the SSP.

This study is primarily based on an intensive empirical field-level research carried out extensively in different locations across the designated command area of SSP. The study also covered the areas which are reaping the benefits of SSP in the form of irrigation through lifting water from the Narmada main canal and branch canals, and drinking water from the Narmada canal-based regional water supply project. The research was carried out to analyze the economic value of these multiple benefits, the direct and the indirect ones: those which range from economic to social and ecological and environmental. It involved a longitudinal study which compared pre- and post-Narmada scenarios of irrigators, farm laborers, and drinking water users.

The study on the direct and indirect impacts of gravity irrigation covered the districts of Narmada, Bharuch, Vadodara, and Panchmahals in Gujarat. The study of the impacts of irrigation through canal lift covered Narmada, Bharuch, Vadodara, Ahmedabad, Mehsana, and Surendranagar districts. The impact analysis of Narmada canal on the well irrigators was undertaken for Ahmedabad, Bharuch, Mehsana, Narmada, Panchmahals, and Vadodara districts. The study on the direct and indirect socioeconomic impacts of drinking water supplies covered rural and urban Bhuj and Jamnagar. The sampling for primary survey to study the impact on farming enterprise was done in such a way that all agro-climatic subregions, over which the design command of SSP extends, are covered.

The analysis of impact of irrigation on farming enterprise goes well beyond the conventional research of agricultural economics and treats the farm as the unit of analysis of private returns, instead of selected individual crops. The analysis covered all the crops grown in all seasons, along with livestock. Inclusion of dairy farming is particularly important due to the fact that it is one of the major subsectors of agriculture in Gujarat, accounting for nearly 20 percent of the agricultural gross domestic product (GDP). Moreover, in addition to the direct private gains which are social and economic in nature, the study captures the positive externalities of interventions including drinking water supplies.

The study assumed great significance in the wake of the recent debate in Gujarat about "miracle growth" in agriculture in the state, which caught international attention. Since the first claim made in a paper published by a group of scholars in 2009, there were a few attempts at validating or contesting these claims (Dholakya and Dutta, 2010; Kumar *et al.*, 2010b). The authors of the original work, which formed the basis for this claim, argued that hundreds of thousands of small water-harvesting structures built by the government in Saurashtra and north Gujarat, the reforms in the power sector, and large-scale promotion of micro-irrigation (MI) systems led to a miracle growth during 2001–2008, with a compounded annual growth rate of 9.6 percent (Shah *et al.*, 2009). Subsequently, this claim was contested by the work of Kumar *et al.* (2010b) which showed that the actual growth rate during the said period was much less and that the phenomenon was a recovery from a major decline in agriculture production due to two consecutive years of drought in 1999 and 2000. Kumar *et al.* (2010) argued that SSP, which had already started irrigating nearly 0.4 million hectares of land by that time, played a crucial role in boosting agricultural production in the state not only by increasing the amount of surface water available for irrigation, but also replenishing the groundwater resources in parts of south, central, and north Gujarat.

The book presents the findings of the above-mentioned research study of SSP on the following: (1) the direct economic impacts of gravity and canal lift irrigation on the farming enterprise, (2) the socioeconomic impacts such as changes in wage employment in rural areas and changes in expenditure pattern of farm households including food consumption expenditure, (3) the indirect economic impacts, that is, the positive externalities of canal irrigation, viz., the impact on the farm income of well irrigators in the command owing to expansion in irrigated area, increase in yield of irrigated crops and dairy animals, economic gain through energy saving benefit for groundwater pumping for irrigation and domestic water supplies owing to reduced depth to water levels, cost saving through reduced well failures, benefit through reduced spending on tanker water supplies owing to improved yield of domestic water wells, and increase in farm wages, (4) the socioeconomic impacts of drinking water supplies, including that on family health, time spent in water collection, and access to water supply sources, (5) the indirect benefits of drinking water provision through Narmada, viz., economic impact of energy saving from reduced dependence on groundwater for water supplies in rural and urban areas, and improved wage earning potential of members of households, and (6) indirect benefit of hydro-power generation, which is cost saving through preventing carbon emission that is likely to be caused by producing electricity using fossil fuel.

The greatest opposition the project faced was on environmental and human rights front (Morse and Berger, 1992; Fisher, 2001; Patel, 2001; Verghese, 2001), as anti-dam activists projected the dam-building exercise as a major catastrophe for ecology of the valley and the indigenous people (Kumar and Ranade, 2004). Hence, the book also discusses the changes in socioeconomic conditions of the tribal families, displaced by the dams and resettled in Gujarat, over the past two decades.

The second chapter sets the global context for the study. The chapter contains an extensive literature survey on the social, economic, and ecological impacts of large irrigation dams in the world. The review covered studies which examined the impacts of large water systems on river ecology, agricultural production, human development, regional economic growth, poverty reduction, and distributional equity in income. The objectives and criteria of evaluating the performance of large water systems in the developing economies are discussed in the context of developing economies. While doing so, it reflects on the scientific validity of claims made by some authors about the future of India's irrigation.

The third chapter provides a brief introduction to the historical necessity of SSP in Gujarat, particularly the importance of the project and its evolution. Some projections of future water demand in various sectors in Gujarat against the potential supplies under two scenarios of water resource development, that is, with and without SSP, are presented. It shows how the project has become crucial to balance the hydraulic equilibrium in a state which is characterized by major regional variations in renewable water availability, but high demands for water in the regions with poor water endowment and low demands in the regions of rich water endowment. The authors use three major indicators used internationally for water scarcity, viz. physical water scarcity index, developed by M. Falkenmark, water scarcity based on demand and supply of water, and water scarcity, which uses the concept of environmental water stress, developed by Paul Raskin, to assess the magnitude of water scarcity problems faced by different regions in the state of Gujarat.

This chapter also discusses the projects' envisaged goals and provides a vivid account of the project area, comprising the environmental, socioeconomic, and other features. The physiographical and agricultural profile of the command areas and the regions which are covered under Narmada canal-based drinking water supply scheme are presented. A macro-level scenario of irrigation, domestic water supply, and power in the state is also presented. A discussion on the physical characteristics of the irrigation and drinking water supply infrastructure built around SSP are also discussed in the chapter.

An outline of research approach and methodology employed and sampling procedure selected for the study is presented in Chapter 4. The chapter also covers the specific methodologies and analytical procedures for the estimation of key output variables such as incremental farm surplus, various technical and economic externalities induced by canal irrigation, and drinking water supplies are discussed in the respective sections of the book for case specific clarity. The methodology used is unique in the sense that it helps actually quantify the positive economic externalities of irrigation, for every unit of additional irrigation produced by the scheme, based on quantification of technical externalities. It also quantifies the indirect benefit from every unit of additional hydropower generated by the scheme owing to preventing carbon emission, which otherwise would have been caused by the production of same amount of energy through fossil fuel.

One of the most significant, yet unintended, benefit of SSP is the improvement in groundwater conditions in the command area owing to recharge from irrigated fields and seepage from canals. Particularly in the context of Gujarat, these benefits are extremely valuable in the view of the fact that groundwater resources are overexploited in the regions which are intended to be covered by the project, due to excessive withdrawal for agriculture. The impacts of canal water introduction on groundwater are analyzed using secondary data on water level available from the periodic monitoring undertaken by SSNNL in the entire command area. The changes in water levels and water quality were analyzed and mapped using a geographic database. The results are discussed in Chapter 5. The positive externalities induced by these physical impacts on groundwater irrigation would be analyzed in the eighth chapter.

It was seen that number of irrigated crops, crop yields, extent of cropped area per farmer, income from crop production, livestock holding, and income from diary production and demand for labor have all increased after the advent of canal water. These direct economic impacts were estimated and presented for all the study districts, covering the farmers using irrigation water by gravity and farmers using water from canal lift. The sixth chapter presents the estimates available from primary data analysis of changes in the cropped and irrigated area, the yield, the cost of inputs, the gross income, and the net income of various crops grown by farmers; change in the yield and income from dairy production; and change in the overall net income from farm after the introduction of Narmada canal water. The outputs are presented for all the five districts where Narmada canal water is directly supplied through gravity and the six districts where farmers are using canal water through lifting. In addition, the impacts of changes in farming enterprise on family consumption expenditure of irrigator families in different districts are also discussed.

One of the land mark impact of the Narmada Project has been on the improved access to drinking water. Safe water for drinking and domestic use has changed the household dynamics. The impact is seen in the form of reduction in drudgery of the womenfolk, in addition to the improvement in the quality of water supplied. The drinking water became potable after Narmada canal supplies.

Prior to the introduction of water supplies through Narmada, the people in Saurashtra and Kutch regions used to consume water from wells which contained fluorides and TDS above permissible levels, while after the introduction of Narmada water, they were saving expenditure on health owing to reduction in the incidence of waterborne diseases in the households. The socioeconomic impacts of drinking water through canals in both rural and urban areas are discussed in Chapter 7.

The impacts include change in the volume of good quality water consumed by households for various domestic uses, reduction in time spent in water collection, reduction in health expenditure, and increased availability of time for seeking wage employment.

Energy saving in agricultural production came as a corollary to the introduction of Narmada canal water. The introduction of canal water directly impacts on groundwater regime in two ways: first, it results in an overall reduction in groundwater pumping. Second, it induces additional recharge in the command area, through irrigation return flows. As a result, there has been improvement in water levels almost everywhere across the command.

Rise in water levels in wells meant a reduced economic cost of energy used for pumping groundwater. This rise in water levels had a significant impact on the farm income of well irrigators as well yields improved and farmers started growing cash crops which needed more water and which gave higher income returns. It was also seen that this has resulted in reducing the dependence on well irrigation altogether. The project certainly contributed to the savings in expenditure by the farmers on well deepening. In sum, there are also several positive externalities accrued to the project area owing to the reduction of well failures, improved sustainability of groundwater-based drinking water sources in the command area, improved sustainability of well irrigation in the command, increase in demand for wage labor in agriculture, and reduced energy requirement for pumping groundwater owing to the rise in water levels.

The pipeline network of Narmada and Mahi canals is planned to serve nearby 54 percent of villages in Gujarat apart from 125 towns. There are yet many villages in south Gujarat which were not brought under this network. The drinking water wells also get benefited from return flows from canal-irrigated fields and canal seepage, improving their sustainability and reducing the dependence of communities on water purchase during summer months. The analyses of these indirect impacts of canal irrigation and drinking water supplies are presented in Chapter 8.

The first section in this chapter quantifies the positive externalities of canal irrigation, namely reduced economic cost of energy used for well irrigation in the region, increased income return for well irrigators owing to improved well yields, reduced investment for deepening of farm wells in the command area, cost saving due to reduced energy use for pumping groundwater and reduced dependence on tanker water purchase for domestic water supply, and improved wage rates for farm and nonfarm laborers in the region, all evaluated through various methodologies.

One of the indirect benefits of Narmada canal-based water supply is the reduced cost of energy used for pumping groundwater in towns and villages. Nearly 24.3 million persons are directly benefited by Narmada canal-based water supply. The second part of the chapter quantifies the economic benefit due to saving in the cost of electricity for pumping groundwater.

The third set of social benefits accrued from the project is related to the price of land. With the irrigation water from the project reaching new areas, the price of land in the drought-prone areas, which are expected to be brought under SSP command, has been on the increase. This section would provide detailed account of how the land value has appreciated over the years in the command area on account of the expected increased economic benefits from irrigation in these areas.

On an average, the SSP generates 3,500 million units of electricity annually. This is clean energy. Chapter 9 presents the methodology for estimating the positive externalities of clean power generation from SSP, which prevents carbon emission that is likely to be caused if conventional fossil fuel-based power generation plants are built to generate equivalent amount of electricity. Here, the cost of mitigating carbon emission from a conventional thermal power plant which produces 1 kW of energy was treated as the social benefit of generating the same amount of electricity through a hydropower plant. The economic value of environmental services from induced river flows is arrived at the travel cost method, where in the cost of travelling to the near site where similar facilities are available is worked out.

The Government of Gujarat has well-thought resettlement and rehabilitation (R&R) policy for project-affected families (PAFs), which any provincial government in India had ever come out with. The policy was formulated in 1987 for successful implementation of SSP, keeping in view the socioeconomic, environmental, and cultural interests of the indigenous communities which were displaced. It marked a major departure from the earlier practice of paying compensation to the oustee families, in which all PAFs, including encroachers, agricultural laborers, and all majors in their family, were given land titles in the new settlements to practice agriculture. This has come about as a result of many years of struggle by the tribal people and a stiff opposition to the state government initial R&R package by the grassroots and civil society organizations working for the cause of indigenous people. So far, the state had successfully resettled 10,065 families from MP and Gujarat, which were displaced by reservoir submergence. Chapter 10 documents the impact of the project interventions on the socioeconomic conditions of these families in Gujarat. This is based on an extensive review of the studies available on the impact of the R&R policies and programs implemented by the government on the project-affected communities at different points of time and the recent field work carried out in some of the new settlements of the project affected people (PAP).

Traditionally, discussions on enhancing the benefits from irrigation schemes had focused on improving utilization of the created potential, through command area development (CAD) activities, including construction of water courses and field channels and drainage network, and formation of water users associations (WUAs) (Kumar and Bassi, 2011). But, Sardar Sarovar is a complex water resource system, which takes irrigation water to different regions in Gujarat with distinctly different agro-climatic conditions and water resource endowments. It is a truism that the marginal returns from irrigation would be relatively low in water-rich regions, but high in water-scarce regions. It means that greater economic and environmental values could be realized by allocating more water to the water-scarce regions of north and central Gujarat, which are also facing problems of groundwater mining. Therefore, along with CAD and formation of WUAs, options such as interregional water allocation should be explored in the context of SSP.

Chapter 11 explores technological and institutional options for maximizing the economic and ecological benefits from SSP in future, without changing the original irrigation water allocation plans of SSP. The options which are explored for maximizing the benefits are: allocating uncommitted and unutilized monsoon flows from Narmada River to water-scarce north and central Gujarat for use during kharif to promote increased recharge of the mined aquifers, through unlined canals; local institutional development and promotion of interregional water trading to encourage economically efficient use of water; and promotion of MI systems for canal water users in the command area.

Chapter 12 summarizes the key findings from chapters in the book. Based on these findings, recommendations are made vis-à-vis the strategies for maximizing the positive economic, social, and ecological impacts and financial viability of large multipurpose projects. Finally, some important inferences are drawn vis-à-vis the objectives and criteria for evaluating the benefits of large multipurpose water resource projects and water resource development policies for developing countries. The authors also argue that if the range of social, economic, and environmental benefits of large water resource systems, which are indirect and intangible, are well articulated and quantified, it would remarkably increase the ability of governments in developing countries to overcome the growing resistance from local communities and environmental groups. Further, effective communication from the governments to the people on the measures being taken up to mitigate negative impacts will also help to generate more support for the large multipurpose water projects.

2

Setting the Global Context

Globally, with growing concern about human rights and environment, the large dams, including Sardar Sarovar, have taken the central stage in the discourse on development and environment (Fisher, 2001). This discourse is characterized by totally opposite views about how far large dams can contribute to development and poverty reduction in the developing countries against the environmental and social problems they pose, or in other words, whether the trade-off between development gains from large water projects and the negative social, ecological, and environmental effects can be balanced (Shah and Kumar, 2008). The anti-dam activists project dam-building exercise as a major catastrophe for the ecology of the Narmada valley and the indigenous people living there (Kumar and Ranade, 2004).

The belief among development activists and practitioner that alternatives to large water systems for providing food, water, and energy security exist (Dharmadhikary, 2005; D'Souza, 2002; Iyer, 2012) has strengthened the resistance to the proposal of investing in large water systems, whereas scholars argue that developing countries have no other choice than to go for large storages. The issue, according to them, is how best to improve their overall effectiveness for human welfare, poverty eradication, and preservation of environment (Biswas and Tortajada, 2001) and how to make those who pay the cost of construction of such large structures the beneficiaries (Tortajada and Biswas, 2012).

In the recent past, a new dimension has been added to this controversy of "politics of dams and development." This is the "politics of hydrology." The main contention of dam opponents is that the design of large dams, including Sardar Sarovar, is based on inadequate hydrological data and that the actual flows likely to be available in future would be much less than what was anticipated, as also the benefits likely to be accrued (Mc Cully, 1996). Their criticism is based on the underlying value that it is almost impossible to predict exactly how much water will flow into a dam based on historical stream flow data for a few years. Mc Cully attributed it to climate change as a result of global warming and geological events. The criticism is grounded on a theory, which is developed on the basis of examples from elsewhere,[1] that the dam-building fraternity around the world has shown a tendency to overestimate the stream flows and underestimate the flood discharges. Even the World Bank's independent review mission, which was supposed to look into the delicate issue of rehabilitation and resettlement in SSP, went beyond its terms of reference and made excursions into the vital issue of the hydrology of Narmada (Morse *et al.*, 1992).

In this context a review of available scientific evidence on the impact of large water resource systems on agricultural production, economic growth, and poverty reduction is very useful to create an informed debate. The review exposes some of the myths about the negative impacts of large water systems based on available empirical work and explores some of the positive externalities these water systems can induce on the society on the social, economic, and ecological fronts. The chapter also synthesizes the available studies and reviews on R&R in large water resource projects, particularly the SSP.

Socioeconomic and Environmental Impacts of Large Water Projects

Socioeconomic impacts comprise any changes to the socioeconomic environment that wholly or partially result from a project activity, whereas, environmental impacts are the externalities on the environment

---

[1] The examples cited are of the 25 large dams built in Thailand, and the huge Buendia-Entrepenas reservoir in central Spain.

and ecosystem that result from a project activity. In case of large water projects, socioeconomic impacts are generally evaluated in terms of changes in: community demographics, human health, livelihood activities, employment, agricultural and nonagricultural income, land use, market effects, and public services (including irrigation and water supply). Environmental impacts are evaluated in terms of changes in hydrological regimes (environmental flows); aquatic, terrestrial, and wetland ecosystems; local climate (including carbon emissions); water and soil quality; and recreational and aesthetic value of nature.

Jacob Granit and Andreas Lindstrom (2009) reviewed the current status of medium to large-scale artificial water storage development with a focus on Africa. The paper also assessed the linkages between water, energy, and food security, and the role of water storage facilities in this nexus. It also reviewed the best practices in water storage development and management infrastructure for building sustainable livelihoods and mitigating climate change impacts. The paper argued that: (1) large-scale water storage plays an increasingly important function as a buffer against rainfall variability in support of economic development and building water security effectively planned and well-built dams bring substantial benefits; (2) large-scale water storage contributes to regional integration and benefit sharing; (3) climate change brings a new dimension to the role of water storage and there potential in developing hydropower to mitigate climate change impacts through wider use of renewable energy.

Investment opportunities with partners in the implementation of the infrastructure component of water storage include the following: Investments can be made in upgrading the existing technology in order to harvest greater power generation from the existing power plants without expanding the actual storage volumes; in projecting feasibility studies toward improved development; and in modernizing irrigation schemes relating to infrastructural components of irrigation networks such as pipelines, pumps, and canals, thus ensuring the functionality of these elements. Water storage technology is another key area that is developing rapidly with run-of-the-river technology gaining acceptance as a way to reduce the need for large-scale water storage. Mitigating environmental impacts is appropriate for development finance and can include project components focusing on watershed management, environmental flows, and other aspects. Investment opportunities in the governance aspects of water storage include the following: institution and capacity building; resettlement; strategic sectoral, social, and environmental assessment; regional integration and power market development to deepen regional integration.

Shah and Kumar (2008) illustrated the role of large storages in development and economic growth, particularly for poor and developing countries; discussed the criteria used by various national and international agencies in defining large dams and identifying their limitations in the context of developing countries; evolved meaningful criteria for defining large storages, which adequately integrates the growing social and environmental concerns associated with dam-building; and identified the gaps in the current cost–benefit calculations to set out new objectives and criteria for evaluating impacts of large dams in developing economies.

The three issues investigated in this study were: (1) role of water in development and growth and the role of large dams in particular; (2) do the current technical criteria in classification of dams as "small" and "large" adequately capture the magnitude of likely negative social and environmental impacts they can cause? and (3) are the objectives, criteria, and parameters currently used to evaluate the costs and benefits of large water impounding and diverting systems, sufficient to make policy choices between conventional dams and other water harvesting systems or groundwater-based irrigation systems?

Analysis of data from 145 countries shows that improvement in water situation of a country determines the progress in human development and economic growth of a country (Kumar, 2010). The sustainable water use index (SWUI), an index derived from water poverty index, and which captures attributes, viz., access to and use of water, water environment, and human resource capacities in water sector, to a great extent, determines the human development, driving economic growth. While the relationship between SWUI and human development index is linear, that between SWUI and per capita GDP is exponential. The paper argued that building large storages would be crucial for improving the overall water situation of a country, as the widely talked about alternatives such as intensive use of groundwater, virtual water trade, and small-scale water harvesting suffer from many limitations. Large dams are important for human development and economic growth.

The study on the basis of empirical data on 14,000 large dams around the world showed that the height of the dam and the storage volume of the reservoir does not reflect the magnitude of negative social and environmental consequences of dams, and, hence, cannot be used as criteria for defining dams as small and large. The study further found that the number of people displaced by dams is a linear function of the total area submerged by them. Every 1 sq. km of the area submerged by large dams in India displaces around 154 people. Using the estimate of 49,660 sq. km as the area submerged by large dams, the total population displaced by them was estimated to be 7.845 million people. The area submerged by dams could be an important criterion for deriving more reliable statistics about displacement; the available estimates of dam-related displacement in India are gross overestimates, in an order of magnitude of 8. Further, large water resource projects produce several positive externalities on the society, such as improving the sustainability of groundwater use through recharge, reducing energy use for groundwater pumping, thereby reducing carbon emission, improving regional and national food security by lowering cereal prices, drinking water security, and drought mitigation. On the basis of these findings, it was argued that the criteria for the evaluation of costs and benefits of dams need to be made more comprehensive taking into account all possible externalities associated with ecological, environmental, economic, and social benefits that dams are expected to accrue.

Matli (2005) used qualitative data to determine the social effects or impacts that the completed water projects in Ha Katse and Ha Mohale have had on the quality of the life of the Basotho in general as perceived by the people. Aspects covered were the quality of life in the social and environmental context and they encompassed arable land, homes, rights, and access, fauna and flora, river flows, infrastructure, energy, economy, and the cultural life. Benefits and opportunities created were determined with the aim of establishing an overall impact on the local villagers and the Basotho on the whole.

The study found that the social impacts of Lesotho Highlands Water Project (LHWP) are manifested in two major categories: environmental and social perspectives. In many countries, a large-scale development project that uses up large areas of land is bound to include the areas inhabited and/or developed by indigenous people. It has been also the case with the LHWP. The decision that governments and concerned development agencies make to proceed with the proposed development on the selected area usually depends on the merits of the very development. The paper argued that it would be ironical to develop an occupied land and yet destroy the natural ecosystem and destabilize the existing communities. Large dam development has to maintain a delicate balance between development on one side and the natural and human environments on the other, as shown by several studies.

The paper argued that the greatest strength of this water project was the economic advantage it has brought to Lesotho. The study found that LHWP boosted the country's revenue earnings through customs tariffs and water sale. The positive effects emanate from a number of notable achievements which could not have been realized were it not for the project, no matter how limited or ill-managed some of them have been. Examples are: (1) the guaranteed earning or harvest from the project compensation scheme; (2) the provision of medical facilities such as clinics and hospitals; (3) the availability of hydroelectric power; (4) the facilitation of agriculture and nonagriculture-based development programs; and (5) improved water supply and sanitation to many communities, and many additional secondary benefits. LHWP's weakness is multifold and rests in its incipient design and social operations. According to the author, perhaps the greatest pitfall of the project was in its initial design. The idea of planning a commercially oriented dam is good where land is not a concern. But to construct it in a place where it takes up a little available agriculturally productive land which can never be replaced except to rely on compensating, it is a matter of grave concern. Large dam developments seem more successful in the official documents of the proponents than on actual communities they are intended to help. This is because projects take up useful land and the proponents persuade the locals to relinquish it by brandishing wonderful development promises which later may not be all fulfilled. The LHWP is cause célèbre with an antithesis of effects and many contentious socio-environmental features.

Skinner *et al.* (2009) analyzed experience and approaches around the world in promoting the sharing of benefits from large dams and made proposals for moving forward on this issue in West Africa. This is best done through a multi-stakeholder partnership of government, industry and civil society interests to maximize the value added through a shared learning approach that provides for wide dissemination of results.

The study found the following ongoing efforts to ensure safeguard in different settings: (1) concrete ways to Integrated Water Resources Management (IWRM) principles that treat water as an economic, social, and environmental good must be found; (2) all stakeholders must work in partnerships to achieve the integration of these elements and dimensions; (3) poverty alleviation must be given an explicit focus on infrastructure provision, especially for large dams that often have disproportionate adverse impact on local communities and traditional river users; (4) cross-sectoral synergies in land management, local income generation, and sustainable management of dams as physical assets must be captured; (5) local actions to protect and manage aquatic ecosystem functions and services in rivers, flood plains, and wetland areas that people rely upon for livelihoods must be funded; (6) innovative measures and incentive mechanisms that build local capacity to adapt land–water resource systems to climate change must be provided; (7) the equitable sharing of benefits is a way of thinking, as well as a practical approach to catalyze and fund local actions that join many strands of water governance reform and sustainable thinking under the IWRM framework, and such mechanisms reinforce social equity in infrastructure strategies and promote sustainability.

Following were the conclusions from the study. Benefit sharing is likely to play an important role in dams and development in West Africa in future. The timing depends on advocacy and how successfully we make the case that equitable sharing of benefits is both a philosophy and an integral part of sustainable development. In multi-stakeholder discussions it is important to keep in mind that nonmonetary forms can be as valuable to rural populations as the monetary forms of benefit sharing. It is not just about sharing revenue, but also about empowering self-reliant community development, ensuring commitments to sustainably manage dams and to unlock the potential of local entrepreneurs to advance new ideas such as payments for ecological services. On monetary aspects, it is important to keep two key questions separate: (1) the source of money for revenue sharing, which is a government economic regulation decision and (2) the mechanisms for the allocation and delivery of benefits to dam-affected and local populations, which is a local development decision.

Ester Duflo and Rohini Pande (2007) examined the productivity and distributional effects of large irrigation reservoirs in India by examining how large reservoirs have affected irrigated and cultivated area and agricultural production both in the districts where they are constructed and downstream, and how the construction of large dams affects rural welfare.

It was observed that the effect on irrigation by the dam in its own district was also positive. Case study evidence (and our model) suggests that reservoir-based irrigation causes farmers to shift toward water-intensive crops. The finding of the study suggests that irrigation reservoirs increase the variance of agricultural production in their own district, without significantly increasing the mean production in the average district. If risk aversion decreases with income, then this increase is likely to be particularly harmful to the poor. As regards rural welfare, reservoirs lead to an insignificant decline in the mean per capita expenditure in the district where they are located and a marginal increase in per capita expenditure in downstream districts. Each dam is associated with a significant poverty increase of 0.77 percent in its own district. In contrast, poverty decreases in downstream districts. Since there are, on average, 1.75 districts downstream of each dam, the poverty reduction in downstream districts is insufficient to compensate for the poverty increase in the dam's own district. The study did not find any evidence to the effect that dams increased the district-level prevalence of waterborne diseases, here malaria. This suggests that negative health effects did not drive the observed increase in poverty.

Malik (2006) addressed the following crucial concerns in the large dam debate: (1) Does development and more efficient management of water resources lead to a reduction in poverty? (2) Empirical analysis is provided to demonstrate that the investments in large water resource development projects, such as multipurpose dams, benefit all sections of society, including the poor and the landless, thereby serving a major mechanism for combating poverty; (3) water resource development and sharing of economic benefits; estimation of multipliers (analytical tools); estimating indirect economic impacts of dams (empirical illustration from Bhakra Multipurpose Dam System); multiplier effects of the

Bhakra Dam in the state of Punjab; Social Accounting Matrices (SAM)-based multiplier model of the Punjab economy; income distribution impacts (estimated from the SAM-based model).

The study found that increased availability of water for irrigation resulted in higher agricultural output. The gross irrigated area during the 40-year period (1996–1997 over 1955–1956) however, increased by about 220 percent, that is, from 3.2 m. ha to 10.3 m. ha. The increase in the irrigated area from the Bhakra system and other sources led to a significant increase in food grain production in Punjab and Haryana; availability of water for industry, household enterprises, and drinking for households and livestock; generation of hydropower; and moderation of floods reducing flood damages significantly (Rangachari, 2005). These changes in availability of additional food grains and electricity in the Bhakra system have led to a number of other benefits in the bene-fited region—100 percent rural electrification in Punjab and Haryana, widespread installations of private wells, a very significant reduction in poverty, overcoming problems of recurrent floods, etc.

The modeling study showed that the Bhakra Dam project generated significant indirect or downstream effects in the state of Punjab. The estimated multiplier value of 1.90 implies that for every rupee (100 paisa) generated directly by the project, another 90 paisa was generated in the region as downstream or indirect effects. These multipliers include the effects of interindustry linkages as well as the consumption-induced effects but do not include the benefits reaped in by other players such as immigrant labor, etc. If it were possible to account for these impacts as well, the value of the multiplier would perhaps be much larger than esti-mated. Further, the investment in the Bhakra Dam has provided income gains to agricultural labor households that are higher than those for the average households.

FAO (2003) carried out a preliminary review of the impact of irriga-tion on poverty with special reference to Asia region. The aim of the study was to provide a framework for analyzing the impacts of irriga-tion on poverty and to review some of the evidence of these impacts. Irrigation affects poverty via a variety of different transmission effects that vary by technology type and by the characteristics of different types of poor. The chief effects are via increased employment and lower food prices: most of the poor (even the rural poor) gain an increasing share of their income from employment and are net food purchasers. Along with raising mean levels of employment, output, and incomes, irrigation can also help reduce the variance of each, although there may be increased covariance. However, the distribution of ownership of and benefit from water and water-yielding assets, for example, between large and small farms, is an important issue. As some of the studies reviewed have suggested, increases in mean yields, output, and incomes are not always replicated across the distribution of farms. Although few project evaluations explicitly address the equity issues of irrigation projects, it is possible to draw a number of tentative conclusions.

The review concluded that irrigation by itself is an important tool in poverty reduction. Regions with the best poverty-reduction performance have greater proportions of irrigated land that has complemented advances in other areas of agricultural production. There are important potential benefits of irrigation such as increased yields, higher and more stable outputs, lower consumer prices, and greater demand for labor that arise solely through the adoption of irrigation but can be magnified when used in combination with other inputs. However, the poverty-reduction impact of irrigation depends much on the detail. Small-scale, low-cost, and labor-intensive irrigation techniques are likely to be more important for poverty reduction. Irrigation techniques that can be accessed by small, capital, or credit constrained farms that use additional labor beyond the initial construction phase are more likely to be of benefit to the poor than large-scale, capital-intensive technologies. But this may not be appropriate for all regions. Substantial poverty reduction in Sub-Saharan Africa is unlikely to be achieved without some new large-scale irrigation projects. The high costs of this, combined with future increasing pressures on water use (for example, subsidized agriculture water use, growing domestic and industrial use), will see big shifts of costly intensive irrigation, from cereals and staples to high-value crops. This, according to the study, required more water control in semiarid areas, and lower cost irrigated areas, for staples' production and employment.

The study further noted that in areas of extreme land inequality such as southern Africa and maybe Latin America, irrigation inequality is even more extreme. Giant farmers have secured free water for capital-intensive use, leaving almost no water control for the labor-intensive small-farm poor. This issue of distributional equity needs to be addressed to achieve poverty-reduction impact of irrigation. Small users and those in tail ends of systems need to be able to secure access to water in the appropriate quantities and at the appropriate times. Water markets and water pricing may be methods of ensuring equitable access, as well as transparent, accountable decision-making institutions. Studies of successes and failures of irrigation in Sub-Saharan Africa show that a combination of supply augmentation (new development of surface water and groundwater, reuse of agricultural drainage water, industrial recycling, waste water use, water harvesting, and desalination) and demand management will be required. Effective demand management will require water resource policies involving in part, cost recovery, transfer of management responsibility, and institutional change. Both the infrastructure and farmer experience to exploit this potential are currently missing in Africa.

Irajpoor and Latif (2011) analyzed the performance of irrigation projects and their impact on poverty reduction and community empowerment in arid environments. The study involved selection of indicators for performance evaluation of irrigation projects, testing of the indicators in the study area, evaluation of results, and guidelines and framework development. The study found significant increase in cropping intensity, crop yield, and irrigated area after the construction of a canal network of the irrigation project. Further, socioeconomic conditions have improved due to development in agriculture as a result of irrigation from the dam. The literacy rate increased to 74 percent in 2006 which was 41 percent before the construction of the dam. Similarly, job opportunities and quality of life increased due to availability of various services leading to reduction in poverty from 15 percent, before construction of the dam, to 7.3 percent by the year 2006. These results clarify that impacts of the dam construction though being significantly positive are not the same as the targeted objectives envisaged in the feasibility report of the project. The results of this study showed that there is only 50–60 percent achievement in some of the planned objectives, while no notable progress was observed in the others. The institutional framework proposed by the study aimed at strengthening the management of water resource projects by involving social and technical organizations.

Singer (2007) examined the influence of large dams on hydrology by examining the drainage network of Sacramento River basin in

California. The study documented basin-wide patterns of hydrograph alteration via statistical and graphical analysis from a network of long-term stream flow gauges located various distances downstream of major dams and confluences in the river basin. For the purpose, stream flow data from 10 gauging stations downstream of major dams were divided into hydrologic series corresponding to the periods before and after dam construction. Pre- and post-dam flows were compared with respect to hydrograph characteristics representing frequency, magnitude, and shape: annual flood peak, annual flow through, annual flood volume, time to flood peak, flood drawdown time, and inter-arrival time. The use of such a suite of characteristics within a statistical and graphical framework allows for generalizing distinct strategies of flood control operation that can be identified without any a priori knowledge of operation rules. Dam operation is highly dependent on the ratio of reservoir capacity to annual flood volume (impounded runoff index). Dams with high values of this index generally completely cut off flood peaks, thus, reducing the time to peak, drawdown time, and annual flood volume. Those with low values conduct early and late flow releases to extend the hydrograph, increasing the time to peak, drawdown time, and annual flood volume.

The analyses showed minimal flood control benefits from foothill dams in the lower Sacramento River (that is, dissipation of the down-valley flood control signal). The lower part of the basin is instead reliant on a weir and bypass system to control lowland flooding.

Malik (2008) made an attempt to demonstrate how the diverse growth-inducing economic impacts of water can be quantified with a reasonable degree of precision. The paper briefly presents results in respect of growth-inducing impacts of investments in the development and management of water resources from three recent case studies. In order to demonstrate how the growth impacts of large-scale multipurpose water resource development projects may differ from small-scale projects, the author analyzed the growth implications of both types of projects. To estimate the regional growth impacts of a large-scale project, Bhakra Multipurpose Dam located in the northwest India and for that of a small-scale project, a check dam constructed in village Bunga in the Shivalik hills region in the state of Haryana were selected. To demonstrate the growth implications of the management of water resources, water-scarce Tamil Nadu was selected.

The results obtained show that, in 1979–1980, the aggregate gross output in the region under the "with project" situation was larger by ₹19 billion (27 percent) than it would have been had the project not been constructed. The project had its biggest impact on the output of agricultural commodities, specially the output of wheat, paddy, cotton, and oilseeds. The output of agricultural commodities was larger by ₹7 billion (61 percent) under "with project" situation than it would have been had the project not been undertaken. The output of electricity was estimated (for 1979–1980) to be higher by 6,033 million units or ₹1,442 million. Further, the value addition estimated using a model shows that in 1979–1980, the aggregate value added in the Punjab economy under the "with project" scenario at ₹42.4 billion was higher than the value added under the "without project" scenario by ₹9.5 billion or by about 29 percent. Of this, the value added from sectors affected directly by the project (agriculture and hydropower) was higher than the corresponding value added under the "without project" scenario by ₹5.0 billion. This shows that the value added in sectors directly affected by the reservoir was almost 49 percent higher under the "with project" situation than the "without project" situation. The value added by sectors indirectly impacted by the dam was higher by ₹4.5 billion (about 20 percent) in the "with project" scenario as compared to the "without project" scenario. Thus, in absolute terms the increase in value added by sectors indirectly impacted by the construction of dam was almost as large as that by the directly impacted sectors.

Urban water demand is growing rapidly in India due to high urban population growth rate and rapid industrialization, a phenomenon sharply observed in many developing economies, including China. Mukherjee et al. (2010) analyzed the role of large surface reservoirs in sustaining municipal water supplies with growing urbanization in India. The study involved: (1) analysis of urban population data for Class I and Class II towns for the period from 1901 to 2001; (2) analysis of data on volumetric water supplies from different sources for 301 towns and cities representing different population size classes collected during the year 1999. The study found that the large urban areas are experiencing faster growth in population as compared to smaller towns. To compound to the problems, most of these large urban centers are located in arid and semiarid regions, which naturally face water scarcity. The study also found that with growing population, the dependence of cities on imported water from large reservoirs becomes higher, as the capacity of local water systems such as streams, rivers, tanks, and lakes and aquifers to meet the urban water demand reduces. The study further found that for large cities, the dependence on surface water import for meeting the urban water demand is as high as 90 percent and above. Further, greater the share of surface water in city water supplies, higher was the average per capita water supply. The study argued that given the structure and pattern of urban population growth, large reservoirs will have a much bigger role in meeting water supply needs in future.

Externalities of Large Water Systems

Canal irrigation is generally considered to be less efficient than groundwater irrigation. But, over the past few decades, there were dramatic changes in the way irrigation efficiencies are analyzed by an irrigation manager (Molle and Turral, 2004). Earlier, irrigation efficiencies were assessed by considering the total amount of water consumed by the crop (evapotranspiration [ET]) in relation to the total amount of water applied in the field. This is based on the notion that the water applied in the field in excess of the crop water requirement is waste (Seckler, 1996; Allen *et al.*, 1998). Such notions no longer help in analyzing the efficiency with which water from irrigation systems is used (Seckler, 1996), owing to the fact that a significant portion of the water applied in the field can actually get captured by downstream users either in the form of well water available for irrigation or return flows (Allen *et al.*, 1998).

Studies carried out in the Indus basin in Pakistan and in the Murray–Darling basin in Australia show that irrigation return flows and seepage from unlined canals are high enough to improve the groundwater regime by raising the water levels in semiarid and arid regions with deep water table conditions if water is introduced for a long time period. Such changes happen due to reduced pumping of groundwater with the introduction of canal water, gradual increase in moisture storage in the unsaturated zone, which increase the unsaturated hydraulic conductivity of the soil media, and the increase in moisture pressure (hydraulic) gradient of the soil (Watt, 2008). While the first factor reduces or keeps the vertical distance for movement of recharge water, the second and third factors increase the rate of vertical movement of soil water.[2] Hence, the "so-called" wastage from canal irrigation produces positive externalities on groundwater irrigation in a region.

As Chakraborty and Umetsu (2003) note, this water can sometimes generate greater economic value than what it would otherwise generate if directly used for irrigation. The reason is that farmers using the recycled water through a lifting device will be able to use it in a more controlled fashion than those who are using surface water irrigation, resulting in better application of water and higher water use efficiency in terms of biomass produced per unit of ET. This can induce a positive externality on the society.

Researchers in the past have highlighted some inherent advantages of well irrigation over surface water irrigation. Daines and Pawar (1987) suggest that well irrigation in India is significantly more productive than surface water irrigation. Similar data are available for Spain (Hernández-Mora et al., 1999). This productivity differences could be due to several factors: better understanding of the resource availability, enabling the farmers to plan their crops; greater access to and control over water enjoyed by well-owning farmers resulting in timely irrigation; and greater control over water delivery from wells resulting in better input use efficiencies and on-farm water management. As illustrated by a study on irrigated rice production in Sone irrigation command in Bihar, timeliness of water delivery and excess water deliveries had significant impact on crop yield, with the impact of the first being positive and that of the second being negative (Meinzen-Dick, 1995).

But, how this positive differential reliability in case of well irrigation does get translated into water productivity gains is a major point of enquiry. Kumar and van Dam (2009) point out two possible reasons for this. First, it is an established fact that while crop yield increases in proportion to increase in transpiration, at higher doses irrigation

---

[2] Based on Richards Equation and van Ganuchten Equation as cited in Watt, 2008, pp. 101–102.

does not result in beneficial transpiration, but non-beneficial evaporation, a component of consumptive fraction (CF) or total water depleted. Irrigation water dosages are normally higher in canal irrigation. This way, increased CF does not result in proportional increase to the yield of crops (Vaux and Pruitt, 1983). Non-recoverable deep percolation is another non-beneficial component of the total water depleted (CF) from the crop land during irrigation (Allen *et al.*, 1998). This also increases at higher dosage of irrigation, which occurs in case of canal irrigation. Thus, the denominator of water productivity will be high in canal irrigation. Moreover, with controlled water delivery, the efficiency of the utilization of fertilizers would be more in the first case. Hence, with improved reliability and water delivery control, both the denominator (CF) and numerator (yield) of the water productivity parameter (kg per m³) could be higher in case of well irrigation.[3]

The second possible reason is that with greater quality and reliability of irrigation in the case of well irrigation, the farmers are able to provide optimum dosage of irrigation to the crop, controlling the non-beneficial evaporation, and non-recoverable deep percolation, with the result that the CF remains low, and the fraction of beneficial evapo-transpiration within the CF or the depleted water remains high. Also, it is possible that with a high-reliability regime of the available supplies, even under scarcity of irrigation water, the farmers can adjust their sowing time such that they are able to provide critical watering. This can bring out high-yield responses. Both result in higher water productivity in kg per ET.

The positive externality, which seepage from canals and return flow from irrigated fields induce on drinking water supplies, is not well documented, except the anecdotes.[4] But, this is highly location specific. The groundwater in the confined aquifers of the alluvial areas of Gujarat contains minerals such as fluoride, and other salts. These make groundwater

---

[3] This can be better understood by the negative correlation between surplus irrigation and crop yields in Sone command that surplus irrigation led to reduced yields (Meinzen-Dick, 1995). Since there are no extra capital investments, it would also lead to higher productivity in economic terms.

[4] The Chief Minister of Gujarat in his recent book *Convenient Action: Gujarat's Response to the Impacts of Climate Change* (McMillan Press, New Delhi) points out several such impacts on health by citing examples from Ahmedabad.

unfit for drinking in many pockets. The examples are districts of north Gujarat which record high TDS levels in groundwater affecting millions of people living in rural and urban areas of these districts, using the groundwater for domestic purpose including drinking and cooking, and Sidhpur and Patan areas which record high fluoride levels in groundwater, affecting the lives of millions of people with serious health impacts in the form of dental and skeletal fluorosis (IRMA/UNICEF, 2001), and negative impact of girls' education. In the rural areas affected by high TDS and fluoride, water had to be supplied from regional water supply schemes which import water from distant reservoirs costing prohibitively to the state exchequer.[5] The large cities, which are facing problems of groundwater quality deterioration, are now depending on water supplied from the Sardar Sarovar reservoir using the Narmada–Mahi pipeline scheme. Dilution of groundwater, occurring as a result of the seepage from Narmada canal and percolation from river beds, would ease the pressure on the regional water supply schemes, and that replaced water could be used for productive purposes.

In the context of large dams rehabilitation and resettlement (R&R) is a crucial issue that attracted activists, researchers as well as policymakers. SSP is one of the most debated, researched, and meticulously planned water projects in the country. Many studies and discussions were held supporting and opposing SSP. Interestingly, both the proponents and the opponents of the SSP used the cost–benefit analysis in support of their arguments. A careful identification of the cost and benefit in items of such a megaproject to society as a whole is a crucial step in the project appraisal exercise. In social cost–benefit analysis, items include costs or benefits of the project to the entire society or part thereof, those should be strictly incremental and ascribable only to the project and also to the effects of the project, both directly and indirectly, on society (Dholakia, 2011).

Although many supporters and Indian government contend that the SSP is a multistate project that caters to drought-prone areas with irrigation and drinking water and has high positive distributional impacts,

---

[5] The regional water supply scheme based on the Dharoi reservoir, which was primarily meant for irrigation, is just one of the many examples of this growing trend.

critics argue that by submerging vast stretches of land, the SSP displaces hundreds of thousands of indigenous people and create environmental problems (Mehta, 2005).

One of the several groups which opposed the project, the NBA, also known as the "Save the Narmada Movement," denounced the SSP for lack of adequate planning. The NBA also pressured the international community to take action to mitigate development-induced displacement which finally resulted in World Bank's withdrawal from funding in 1993. The Indian government, however, opted to continue without the bank funding despite opposition within and outside the country. The Indian government's decision to pursue the project without the bank's involvement forced the SSP to mobilize its forces at the domestic level. In 1994, the NBA filed a public interest litigation petition 21 with the Supreme Court of India. The Indian government halted work on the dam in 1995 because of a Supreme Court ruling recognizing that the rehabilitation of displaced people was inadequate. However, on October 18, 2000, the Indian Supreme Court on a fresh examination of the issue allowed construction of the SSP to be carried out. Although Indian government assured displaced population of SSP with several policies and rehabilitation packages still it was criticized for its shortcomings in basic policy formulation and sincerity in implementation.

The independent review conducted by Morse and Berger (1992) stated that

the World Bank and India both failed to carryout adequate assessments of human impacts of SSP. Many of difficulties that have beset implementation of the projects have their origin in this failure. It failure to consult the people has resulted in opposition to the projects, on the part of potentially affected people, supported by activists. It did not taken account of interests of tribal people. The Bank failed to consider the effects of the Projects on people living downstream of the dam. As a result of both the inadequate database and the failure to incorporate provisions of the Bank's policies in the 1985 credit and loan agreements, the provisions for R&R do not adequately address the real needs of those affected. In particular, the agreements allowed a distinction between "landed" and "landless" oustees which failed to recognize the realities of life in the submergence villages. Similarly, the rights of encroachers were not acknowledged. Significant discrepancies in the hydrological data and analyses indicate that the SSPs will not perform as planned either with or without the upstream Narmada

Sagar Projects. A realistic operational analysis of the Projects upon which to base an impact assessment has not been done.

Anil Patel and Ambrish Mehta (2011) strongly condemned the views of Morse and Berger and said views of the team were anti-SSP, based on distortions. They said SSP suffers from grave methodological flaws that include those which did not follow from factual premises; conclusions based on an inadequate database, errors in reasoning, or failure to carry logical analysis far enough; and sources not properly or adequately cited.

The United Nations' Guiding Principles on internal displacement set out in 1998, the International Covenant on Economic Social and Cultural Rights (ICESCR), and the 1957 International Labor Organization (ILO) Convention 107 assert that project authorities should explore all feasible alternatives in order to avoid displacement altogether, and in situations in which displacement is unavoidable. ILO and ICESCR specifically state for legal rights and interests of indigenous people, respectively. It stresses the need for concerned authorities to minimize the adverse effects of the displacement. But there are no separate Indian laws pertaining specifically to the state's legal responsibility to its internally displaced. Internally displaced population must turn on fundamental rights, writ jurisdiction of courts for recourse. Articles 21, 39, and 41 of the constitution provide a framework for providing the right to life, adequate means of livelihood, and states responsibility in giving right to work, respectively. Provisions of the Land Acquisition Act (LAA), which the government implement for large projects such as the SSP, unfortunately nullify the impact of these landmark decisions and provisions.

In light of the importance of the SSP and the magnitude of its impact on local populations, the Indian government established the Narmada Water Disputes Tribunal Award (NWDT Award) in 1979. The NWDT Award sets the policy framework for R&R. It specifies that resettlement must precede submergence by at least 12 months. However, the NWDT Award does not provide for tribal populations which cannot claim legal title to land despite their customary rights to the land and their cultivation of the land for generations. In addition, the NWDT Award does not mention the status of those who are landless. Although there was establishment of the Grievance Redressal Authority (GRA), Special

Rehabilitation Packages, and clear directions for protection of social, economic, and cultural interests of the displaced people of the Narmada Project, NWDT could not meet the interests of displaced people.

According to Robinson (2003), evidence suggested that for a vast majority of indigenous/tribal peoples displaced by big projects, the experience has been extremely negative in cultural, economic, and health terms. The outcomes have included asset loss, unemployment, debt-bondage, hunger, and cultural disintegration. For both indigenous and nonindigenous communities, studies show that displacement has disproportionately impacted on women and children.

A similar kind of opinion expressed in a fact finding report by L. C. Jain and S. C. Behar (2008) after hearing the problems of *adivasis* (tribes) from Gujarat, Maharashtra, and MP was that the quality of land allotted to many PAFs is extremely poor and consists of uncultivable land. The R&R sites were supposed to be located in the command area of the dam. But this has not happened in many cases, and these sites are mostly without irrigation facilities. Even when the land is close to the water, a common problem is poor or no access to the water, so the people settled there are effectively without water. This is in total violation of the Tribunal's directives. Infrastructure in the *vasahat*s (locality) is poor. There is no response to the complaints or grievances. The administration is completely indifferent, not just to people's complaints, but even to the directives of the GRA, on the occasions when there are any. Serious problems persist, even 20 years after the first oustees were resettled in Gujarat, in recognizing and allotting land to eligible PAFs. The loss of the sense of community is deeply felt and they become vulnerable to tensions and hostility vis-à-vis the host community. They are frequently dependent on the host community even for water. People who owned land earlier have now become landless laborers. Overall, the quality of life has clearly deteriorated for these PAFs after R&R, despite the numerous efforts undertaken.

***

In nutshell, the reviews of literature in global context point the following. Irrigation is a tool for poverty reduction. There are important potential benefits of irrigation such as increased crop yields, higher and more stable outputs, lower consumer prices, and greater demand for labor that arise solely through the adoption of irrigation but can be magnified when used in combination with other inputs. However, the poverty-reduction impact of irrigation depends much on the technology used, the economic structure, and the quality of institutions engaged. Capital-intensive irrigation based on large storages and conveyance systems may not be always the best option, and instead low-cost labor-intensive technology might prove to be more suitable in conditions of labor surplus and small holdings. The flood control benefits from dams would also be determined by the exact location of the dam in the basin. Provision of irrigation water to an area, which is characterized by great inequity in access to land, would only increase the economic inequality.

In the context of developing countries, the main issue concerning water resource development is not whether large dams have an important role to play in the coming decades, since there is really no other choice, but rather how best we can continue to improve their overall effectiveness for human welfare, eradicate poverty and preserve the environment (Biswas and Tortajada, 2001). As Tortajada (2002) points out, "No large-scale infrastructure development project is possible anywhere in the world that only induces positive impacts." Even though some negative social and environmental impacts are unavoidable, they should be looked at in an organized way so that they can be mitigated properly (Tortajada, 2002).

It is equally essential to ensure that those who may have to pay the costs for the construction of these structures (for example, people who have to be resettled or whose livelihoods may be threatened) are made direct beneficiaries of these projects (Tortajada and Biswas, 2012). "Benefit sharing" is likely to play an important role in dams and development in developing countries. The success in building dams depends on how successfully we make the case that equitable sharing of benefits is both a philosophy and an integral part of sustainable development. In benefit sharing among rural populations, nonmonetary forms can be as valuable as the monetary forms. The success also lies in obtaining commitments to sustainably manage dams and to unlock the potential of local entrepreneurs to advance new ideas such as payments for ecological services. The two issues in revenue sharing are: (1) the source of money for revenue sharing and (2) the mechanisms for the allocation and delivery of benefits to dam-affected and local population.

Institutions that favor socially, economically, and politically powerful groups can result in unequal distribution of costs and benefits of large dam construction, thereby increasing poverty of dam-affected people. But large water resource systems such as multipurpose dams benefit all sections of the society and their indirect benefits are often as large as the direct benefits, as shown by a study of the multiple effect of 1.9 for Bhakra Project in Punjab, India. In groundwater overexploited semiarid and arid regions, gravity irrigation can induce several positive externalities by augmenting groundwater recharge, raising the water table, and improving the natural quality of water in the underground formations through dilution, though some of the benefits would be highly location specific. Further, surface reservoirs play a great role in sustaining urban water supplies in large cities of India, with fast-growing urban water demand and the inadequate supplies from local water resources to meet this demand. As cities grow, a greater proportion of the urban water demand will have to be met from import of water from distantly located large reservoirs.

The criteria for defining large water storages need to be changed to capture the magnitude of actual negative social and environmental consequences of large water storage projects such as area of submergence and people displaced. Simultaneously, the criteria for evaluating the benefits and cost of large water storages in developing economies need to be broadened to capture the various positive externalities such as arresting groundwater depletion through recharge, improving regional and national food security through lowering of food prices, reducing energy use for groundwater pumping through raising water table, thereby reducing carbon emissions, provision of drinking water security, and drought mitigation.

Finally, R&R poses a difficult problem especially when the project involves different states and tribal ethnic groups living in uplands and forests. There is need to follow the current international norms. Existing Indian laws were often found inadequate.

To minimize the adverse impacts on displaced population, it is increasingly felt that resettlement must ensure equal rights to women, children, indigenous populations, and other vulnerable groups, including the right to property ownership and access to resources. The following steps were also recommended for minimizing adverse impacts: (1) protect the right to an adequate standard of living and promote economic opportunities; (2) support access to education for displaced children; (3) support steps to guarantee property rights and civil rights; (4) design assistance and protection measures to ensure impartiality; and (5) support, technically and financially, attempts by cognizant authorities to fulfill their responsibilities to the displaced.

3

Social Benefits and Impacts:
An Analysis of the Sardar Sarovar Project

There are not many examples from around the world in which an impact assessment of a mega water resource project is undertaken while the project is under the stage of implementation. The SSP is still under implementation and less than one-third of the design command area is ready for receiving water. Though the Narmada canal-based water supply scheme is on the verge of completion, it was never planned as a component of the project. Instead, the Sardar Sarovar reservoir was only expected to allocate a share of its water for drinking and domestic uses in the parched regions of the state as per a state government policy. For a mammoth project like the Sardar Sarovar, it is important to undertake a midcourse evaluation of the impacts, due to the reason that several questions were raised about the technical feasibility of the project and its likely economic and ecological impacts at the planning stage by its critiques, and a post-project evaluation would not provide sufficient opportunity to the managers for corrective measures in case needed to optimize the project benefits and to minimize the costs.

Some of the recent studies on agriculture and water resources in Gujarat suggested the need for undertaking such a study. There were four important changes happening in Gujarat's water resource landscape. (1) A large additional area was being brought under surface irrigation in a very short span of time, when the contribution of surface water to the state's irrigation came down to almost 20 percent and farmers' dependence on deep tube well irrigation became excessive. (2) More than 6,000 villages and almost a hundred towns in the state began to receive water from pipeline schemes which tap water entirely from Sardar Sarovar Main Canal. (3) Many seasonal rivers of central and north Gujarat started receiving imported water from Sardar Sarovar through its main canal, which transect these rivers. (4) A remarkable rise in groundwater regime in areas which receive water from Narmada either through gravity irrigation or through release in rivers was experienced with dilution of minerals present in groundwater. The project authorities had appreciated the need for undertaking such a study.

Hence, the research study was contemplated to undertake a comprehensive evaluation of the current and likely future impacts of the Sardar Sarovar Narmada Project in Gujarat on the social, economic, and ecological/environmental fronts (Figure 3.1). The specific objectives were to: evaluate the direct economic and socioeconomic impacts of Narmada water supplies through canals in command areas and outside; evaluate the socioeconomic impacts of the introduction of Narmada waters for drinking water supplies in rural and urban areas of Gujarat; evaluate the physical and environmental impacts of introducing irrigation water in the command areas; and carry out an analysis of the real economic benefits of the SSP, which, along with the direct economic costs and benefits, include the positive and negative externalities induced by the use of Narmada water for irrigation and drinking water supplies and hydropower generation from the SSP.

The Approach

Tools in environmental and ecological economics were used to evaluate the economic value of the social benefits (positive welfare effects) (van den Bergh, 2007) and ecological benefits from the project (Kay *et al.*, 1997), respectively. Participatory rural appraisal (PRA) tools and case study methodologies were employed to capture the qualitative aspects of socioeconomic changes (at the household level) that have happened with the introduction of water for irrigation and drinking water supplies.

Figure 3.1:
A Flowchart Depicting the Direct and Indirect Impacts of the SSP

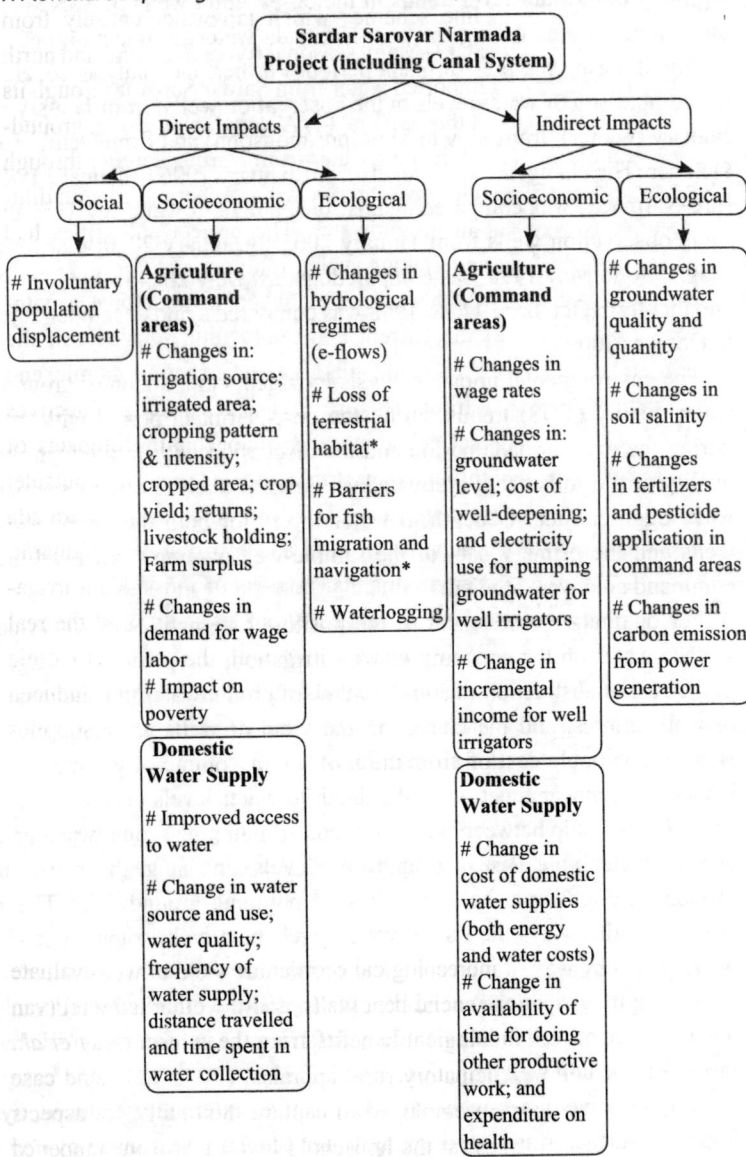

Source: Author.
Note: Changes in items marked as * were not evaluated in the study.

The physical (geo-hydrological) impacts were analyzed by comparing the groundwater level trends in the observation wells of SSNNL in the command area during the post-Narmada water introduction period with that during the pre-Narmada period. Further, the analysis covered the comparison of water levels in the observation wells, from January to January (winter), from May to May (pre-monsoon) and from October to October. The change in water levels from January 1996 to January 1999 (pre-Narmada) was compared against the change in water levels in the same observation wells from January 2005 to January 2010; that from May 1996 to May 1998 was compared against May 2004 to May 2009; and that from October 1996 to 1998 was compared against October 2004 to October 2009.

The environmental impact analysis included: comparison of groundwater quality (TDS) trends during the post-Narmada period with that during the pre-Narmada period and the discussion with farmers on the quality of the soils being cultivated. In order to analyze effects such as waterlogging, data of "depth to water level" for both the observation wells and the primary data of the sample-well-owning farmers in the command area were studied.

For qualitative assessment of the positive externalities of imported surface water on the economy of well irrigation, the potential changes in the cost of abstraction of a unit volume of groundwater, the incidence of well failures, and the change in the yield of wells are considered. Here, the variable cost of abstraction of a unit volume of groundwater is taken as a linear function of the depth to water levels because of the linear relationship between the energy cost of pumping groundwater and depth to water table. Rise in groundwater levels can change the irrigation economy by reducing the energy cost of pumping groundwater. There are many other ways the rise in water levels help well irrigation economy. One is by way of increasing the sustainable yield of wells, thereby increasing the irrigation potential of wells, and the other is by reducing the incidence of well failures, thereby deferring the investments made by farmers to sustain well irrigation.

A qualitative assessment of the positive externality induced by imported surface water on drinking water economy was made by considering the potential reduction in the cost of production and the supply of water for domestic use in rural and urban areas by public utilities, consequent to the increased availability of groundwater in the local aquifer and its quality amelioration, induced by canal irrigation. Here, it is assumed that dilution of groundwater would make it potable, or at least reduce the capital investments required for treating groundwater for potability, in treatment plants, thereby reducing the cost of production of water from underground sources. Also, the augmented recharge of local aquifers would result in increased availability of water in the wells, thereby reducing the dependence on tanker water supply. But, such indirect benefits get affected only in areas which are currently dependent on bore well-based water supply schemes. However, their number is reducing over a period of time because of Narmada canal-based drinking water supply schemes.

An extensive review of the available literature on geo-hydrology and groundwater resources was carried out to ascertain the groundwater recharge and storage potential in north and central Gujarat aquifers. Scientific literature in the international arena dealing with the physics and economics of "dynamics of interaction" between "canals and groundwater system" and "irrigation return flows and groundwater system" in similar geo-hydrological environment was extensively utilized to analyze the nature of impact that SSP waters would have on the geo-hydrology, geo-hydro chemistry, and irrigation and drinking water economy of different regions of Gujarat.

The socioeconomic impacts of the introduction of irrigation water through canals and river channels and new wells were assessed by comparing the scenarios before and after the introduction of canal water with respect to: gross irrigated area, crop yields (kg per ha), livestock holding under different categories, total agricultural output (in ton per annum for different crops), dairy outputs, number of fish ponds and fishery outputs (ton), surplus value product from farming including dairying and fisheries; increase in wage employment gained by farm laborers (labor days and in cash); and reduction in the time spent in activities such as washing, bathing, and animal feeding.

The socioeconomic impacts of drinking water supplies were assessed by comparing before and after scenarios on per capita water use for domestic purposes in different seasons; time spent by women in water collection; children's attendance in schools and school dropout rates; and health expenditure due to water-related diseases. Because of the absence of any prior benchmark studies some reliance on memory with adequate probing became necessary in the field interviews to get pre-project data. The evaluation of social benefits takes the sum of private benefit and positive externalities. The positive externalities were assessed in terms of the following:

- The economic value of the energy saving benefit resulting from rise in groundwater levels. This was estimated by multiplying the full cost of producing one unit of energy and number of units of energy saved for groundwater pumping per ha of well-irrigated area.
- The cost saving, owing to a reduction in the incidence of well failures or well deepening, owing to greater recharge. This was qualitatively assessed by analyzing the reduction in the incidence of well failures in the command area villages per year.
- Increase in income gain from wage employment resulting from increase in daily wage rates for farm laborers.
- The incremental income of well irrigators in the command area resulting from increased well outputs. For calculating this, the average increase in command area per ha of (pre-Narmada) well-irrigated area and incremental net income from crops and livestock per ha of well irrigation were estimated.
- The reduction in the cost of production and supply of water for domestic use in rural and urban area by public utilities, owing to the increased availability of groundwater in the local aquifer and its quality amelioration. This was worked out by multiplying the reduction in the average cost of production and supply of a unit volume of water round the year with the introduction of canal water and the total volume of water supplied.

The private benefit from irrigation was estimated by taking the average incremental net farm income of individual farmers due to the increased availability of irrigation over the pre-Narmada scenario per unit cultivated area, including that from dairying and fisheries. This will be done separately for canal-irrigated area, area irrigated by direct lifting from canals, and area irrigated by river lifting. Obviously, the analysis should take into consideration any changes in the cropping system that have occurred due to the introduction of canal water.

The social benefit of drinking water supplies through the Narmada pipeline scheme was estimated by adding up the positive external-ity induced by drinking water supplies and the economic value of the social wellbeing resulting from increased water supplies. Drinking water supply being a "social good," the economic value of the social wellbeing is estimated using "proxy method" as the cost of producing equivalent amount of water of same quality in the region using an alternative method. The positive externality was expressed as:

- The economic value of the time saved in fetching water for domes-tic uses by men and women. This was estimated in terms of the total number of women/men benefitted by the improved access to drinking water supplies in rural and urban areas in terms of time saving, and the proportion of these men and women who are able to allocate this time for any activity which has economic value.

- The reduction in energy use for pumping groundwater in the cities and villages previously served by water supply from bore wells estimated from the reduction in energy use or electricity charges for pumping a unit volume of groundwater for water supply and the volume of groundwater supply, which is replaced by Narmada canal-based water supply.

The direct economic benefit from power generation was estimated by multiplying the cost of production of 1 kW of energy from an alter-nate source, and the quantum of power currently generated by the plant per year.

Another important dimension of the hydropower benefit is that it is eco-friendly (Joshi and Kapadia, 2010; Shah and Kumar, 2008). A kilo-watt hour of energy produced from hydropower plant gives an additional benefit equal to the cost of environmental damage which a thermal plant would cause for the same amount of power generation (Shah and Kumar, 2008). The positive externality induced by hydropower from the SSP can be equated to the economic cost of carbon emission, which otherwise would occur with a fossil fuel-based power plant, for the same level of electricity that is produced from the SSP today.

In order to evaluate the economic value of the ecological benefits derived from the presence of freshwater in many of the otherwise dry river courses and the emergence of new tree cover in villages and towns, the "travel cost method" was used (Kay *et al.*, 1997).

The changes induced by displacement on the socioeconomic and cultural profiles of the dam oustees in the new settlements were analyzed by reviewing the findings of the surveys carried out by scholars at points of time on their socioeconomic profile, which included one or more of the following: pre- and post-scenarios vis-à-vis farming enterprises, coping strategies for droughts, access to education, access to water supply and public health care systems, access to market, cultural life, overall level of satisfaction in life in the new settlement, and comparing the socioeconomic dynamic of the new settlers with that of the original inhabitants of the new settlements. In addition to drawing on these studies, a primary survey was carried out by the researchers in early 2013 to see the latest situation vis-à-vis the displaced communities.

Analytical Procedure for Estimation of Income from Crop and Dairy Production

Incremental income from crop production for crop i ($£_i$) due to introduction of irrigation water is estimated as

$$ICROP_{POST,i} - ICROP_{PRE,i}, \tag{1}$$

where $ICROP$ is the net income from crop production per unit area, and suffixes POST,i and PRE,i represent post- and pre-Narmada, respectively, for crop, i. The values of variables, $ICROP$ is obtained by subtracting the input cost per unit area from the gross return from crop production per unit area. The gross income from crop production is estimated by taking the yield of the main and by-product of the crop (kg) and the farm gate price per kg.

Incremental income from livestock production for livestock category j($¥_j$):

$$ILSTOCK_{POST,j} - ILSTOCK_{PRE,j}, \tag{2}$$

where *ILSTOCK* is the net income from livestock production per unit of livestock, and suffixes POST,*j* and PRE,*j* represent post- and pre-Narmada, respectively, for livestock type, *j*. The values of variables, $ILSTOCK_{POST,j}$ and $ILSTOCK_{PRE,j}$ are obtained by subtracting the input cost per animal unit from the gross income per animal unit. The gross income from dairy animals per livestock unit is estimated by taking the average annual milk yield from animals per livestock unit and the selling price of milk.

Now, let us assume that the farmers grow a total of *m* crops. Then the average incremental income (or also increase in average income) from crop production per unit of land can be estimated ($\emptyset^1$) as

$$\emptyset^1 = \left\{ \frac{\left[\sum_{i=1}^{m} A_i^1 \times ICROP_{POST,i} - \sum_{i=1}^{m} A_i \times ICROP_{PRE,i} \right]}{\sum_{i=1}^{m} A_i^1} \right\}. \tag{3}$$

Here, A_i is the area under crop *i* pre-Narmada and A_i^1 is the area under the crop *i* post-Narmada water introduction.

Now, let us assume that the farmers keep a total of *n* types of livestock. N_j is the total number of livestock in the *j*th category pre-Narmada water, and N_i^1 is the total number of livestock in the same category post-Narmada. Then, the average incremental income from dairy production per unit of livestock (β^1) can be estimated as

$$\beta^1 = \left\{ \frac{\left[\sum_{j=1}^{n} N_j^1 \times ILSTOCK_{POST,j} - \sum_{j=1}^{n} N_j \times ILSTOCK_{PRE,j} \right]}{\sum_{j=1}^{n} N_j^1} \right\}. \tag{4}$$

The average income from crop production pre-Narmada ($\emptyset$) can be estimated as

$$\emptyset = \sum_{i=1}^{m} A_i \times ICROP_{PRE,i}. \tag{5}$$

Likewise, the average income from livestock production pre-Narmada (β) can be estimated as

$$\beta = \sum\nolimits_{j=1}^{n} N_j \times ILSTOCK_{PRE,j}. \tag{6}$$

The specific methodologies and analytical procedures for estimation of the key output variables such as farm surplus, and various externalities induced by canal irrigation and drinking water supplies were worked out and discussed in the subsequent sections.

Estimation of Incremental Farm Surplus per Unit Area Cultivated in Canal-irrigated Area

Incremental farm surplus ($FARM_{SURPLUS}$) due to irrigation was estimated by using the following variables: average increase in net income from crop production per ha of the gross (sample) cropped area $\varnothing^1$; current gross cropped area in the Narmada command from canal irrigation and lift (A_{GROSS}); increase in gross cropped area in the Narmada command (ΔA_{GROSS}); the average income per ha of gross cropped area prior to Narmada ($\varnothing$); the average holding of different types of livestock per ha of the gross cropped area of the (sample) farmers at present (N); the average increase in the net income from livestock production covering all types of livestock (β^1); increase in livestock holding of farmers per ha of gross cropped area (ΔN); and the average income per unit of livestock (β) as

$$FARM_{SURPLUS} = A_{GROSS} \times \varnothing^1 + \Delta A_{GROSS} \times \varnothing + \\ N \times A_{GROSS} \times \beta^1 + A_{GROSS} \times \Delta N \times \beta. \tag{7}$$

Now, for a unit irrigated area, the farm surplus can be defined as

$$FARM_{SURPLUS-UNIT} = \varnothing^1 + \{\Delta A_{GROSS} / A_{GROSS}\} \times \varnothing + \\ N \times \beta^1 + \Delta N \times \beta. \tag{8}$$

Now, $\left\{\dfrac{\Delta A_{GROSS}}{A_{GROSS}}\right\}$ can be treated as same as the incremental area cropped by the sample farmers against their average gross cropped area, if they are a representative of the canal-irrigated area. This can be further split into average incremental farm income for canal-irrigated areas and areas receiving canal water through lift.

Estimation of Positive Externalities of Canal Irrigation on Well Irrigation

The economic value of the positive externality of canal irrigation on groundwater was estimated using the following output variables, the current gross groundwater-irrigated area in the command area of canals and areas receiving irrigation through canal lift (A_{WELL}); the average increase in the net income from well irrigation per unit of the gross well-irrigated area ($\varnothing^{1}A_{WELL}$); the increase in groundwater-irrigated area post-canal water introduction (ΔA_{WELL}); the net return from groundwater irrigation pre-canal water introduction ($\varnothing A_{WELL}$); the current average livestock population per ha of gross well-irrigated area (N_{WELL}); the increase in the average net income from livestock production of well irrigators per unit of livestock for all types of livestock owned by the well owners (β^{1}_{WELL}); and the average increase in livestock population per ha of gross well-irrigated area (N^{1}_{WELL}) as

$$
\begin{aligned}
WELL_{EXTERN} = A_{WELL} \times \varnothing^{1}_{WELL} + \Delta A_{WELL} \times \varnothing_{WELL} + \\
N_{WELL} \times A_{WELL} \times \beta^{1}_{WELL} + \Delta N_{WELL} \times A_{WELL} \times \beta_{WELL}.
\end{aligned} \quad (9)
$$

The same for a unit of gross well-irrigated area can be expressed as

$$
\begin{aligned}
WELL_{EXTERN-UNIT} = \varnothing^{1}_{WELL} + \{\Delta A_{WELL} / A_{WELL}\} \times \varnothing_{WELL} + \\
N_{WELL} \times \beta^{1}_{WELL} + \Delta N_{WELL} \times \beta_{WELL}.
\end{aligned} \quad (10)
$$

Estimation of the Positive Externality of Energy Saving Benefits in Well Irrigation Due to Canals

The economic equivalent of the energy saving benefit owing to a rise in water levels in wells resulting from return flow from canal irrigation is equal to

$$\text{ECOBEN}_{\text{SAVE - ENERGY}} = \propto \times A_{\text{WELL - IRRIGATION}} \times \text{COST}_{\text{ENERGY}}, \qquad (11)$$

where $\propto$ the total energy saved per annum per ha of well irrigation in kilowatt hour, $\text{COST}_{\text{ENERGY}}$ is the cost of producing and supplying energy, and $A_{\text{WELL - IRRIGATION}}$ is the area under well irrigation. The $\text{COST}_{\text{ENERGY}}$ includes the direct economic cost of producing and supplying electricity and the cost of reducing carbon emission from power generation, which is equal to ₹0.47 per kilowatt hour of electricity.

Positive Externality of Drinking Water Supply Benefits Due to Canal Irrigation

The externalities induced by canal irrigation on drinking water supply schemes include, reduction in energy cost of pumping groundwater due to rise in water levels and improved sustainability of drinking water supply can be computed as

$$P_{\text{WELL}} (R_{\text{P - DEPTH}} \times 0.055 \times V_{\text{DRINK - WATER}} \times \text{COST}_{\text{ELECT}} + V_{\text{DRINK - SUMMER}} \times \text{PRICE}_{\text{TANKER}}). \qquad (12)$$

Here, $R_{\text{P - DEPTH}}$ is as defined earlier for equation (2). $V_{\text{DRINK - WATER}}$ is the per capita volume of water used for drinking in a year in m³; P_{WELL} is the total size of the population dependent on groundwater-based schemes for domestic water supply in the region; $\text{COST}_{\text{ENERGY}}$ is the average cost of generating and supplying 1 kW of electricity; $V_{\text{DRINK - WATER}}$ is the per capita volume of water used for drinking and domestic uses in summer; and $\text{PRICE}_{\text{TANKER}}$ is the market price of a cubic metre of tanker water supplied.

Positive Externality of Energy Saving Benefit Due to Introduction of Surface Water for Drinking

The positive externality associated with the replacement of well water with surface water for drinking, in terms of saving in the cost of electricity (∂) in ₹ per annum could be estimated as

$$\partial = V_{DRINK} \times 10^6 \times (COST - SAVING_{ELECTRICITY}). \tag{13}$$

Here, V_{DRINK} is the water requirement for drinking in the service areas of the drinking water supply grid in million cubic meters (MCM). Here, it is 490 MCM per annum. $COST - SAVING_{ELECTRICITY}$ is the average saving in the cost of electricity used for supplying drinking water per m^3 of water.

Positive Externality of Clean Energy Production from Hydropower

In order to estimate the economic value of this benefit, we follow the following methodology. Every 1 kWh of electricity produced through thermal power stations generates about 0.87–0.96 kg of CO_2 a greenhouse gas, depending on the type of fuel used. In the case of coal-based thermal power, the CO_2 emission is 0.96 kg per unit (kilowatt hour) of electricity, whereas in the case of petroleum-based thermal power, the CO_2 emission is 0.87 kg per unit of electricity. The avoidance of carbon footprint from generation of every kilowatt hour of electricity produced from hydropower is 0.96 kg of CO_2 or 0.26 kg of carbon.[1]

The cost of mitigating carbon emission from a conventional thermal power plant, which produces 1 kWh of electricity, can be treated as the social benefit of generating the same amount of electricity through a hydroelectric plant. David and Herzog from Massachusetts Institute of

[1] These estimates are based on data available from United States Department of Energy on the amount of electricity produced from various fossil fuel based power plants.

Technology had estimated the mitigation cost of carbon emission from various thermal power plants. They estimate that for a pulverized coal power plant, it costs US $49 for reducing carbon emission by one ton. Here, preventing 0.96 kg of CO_2 emission from a thermal power plant would require US $0.047, or ₹0.47.[2] This means, every kilowatt hour of hydropower generated to meet an existing demand for electricity and which can defer the investment in coal-based thermal power plant generates a social welfare effect equivalent to 47 paise, that is, ₹0.47.

The welfare benefit, accrued, therefore, from hydroelectric power generation every year (₹) is:

$$\text{ECOBEN}_{\text{HYDRO}} = \infty \times 0.47 \times 10^{\wedge}6 \tag{14}$$

Here, "∞" is the amount of electricity produced per annum currently in million units.

Type and Sources of Data, and Sampling Procedure

Two types of data were used: primary and secondary. The types of primary data collected from those benefited by water release through canals and rivers are: (1) the crop and livestock inputs and outputs of sample farmers in the command area and in the area irrigated through canal/river water lifting, and their cropping pattern; (2) the drinking water supply situation in the villages affected by canal water release for irrigation, particularly the operating and maintenance cost and water supply levels across seasons; (3) household socioeconomic dynamic in these villages, with regard to drinking water collection; (4) energy requirement for well irrigation in the canal-affected areas; wage labor for agriculture in these villages; and (5) willingness of people living on banks of the rivers to pay for environmental services. All these were collected for both pre- and post-Narmada situations.

---

[2] Here, we have considered purchasing power parity adjusted conversion ratio of US $1 = ₹10.

The types of primary data collected from villages/towns and individual households receiving Narmada water supplies for drinking are: (1) domestic water-use levels in the sample rural and urban households in different seasons; (2) change in hygiene practices of members of the households; (3) the time spent on drinking water collection; (4) school attendance of children; the wage labor earnings of adult members of households, particularly those who are engaged in water collection; (5) annual health expenditure toward treatment of water-related diseases such as fluorosis and kidney stone; and (6) tree plantation activities undertaken by village panchayat and households. Besides these, qualitative data, especially on household socioeconomic dynamic, were collected using interviews and PRA tools.

The secondary data collected include: (1) the total volume of water released from the Narmada system to the command area annually and the area irrigated there; (2) the amount of water released in north and central Gujarat rivers; (3) electricity generated from the hydropower plants annually; the approximate area irrigated through direct canal lifting; (4) time series data on groundwater levels and water quality (TDS, fluoride, nitrate) in the command area and outside; (5) number of villages and towns receiving Narmada water for drinking/municipal water supplies; and, (6) the average annual energy subsidy in agriculture per kilowatt hour. Time series data on food production from districts receiving Narmada waters for irrigation (through canals and direct lifting) was also collected. The data on groundwater levels was supplied by SSNNL.

We used multistage stratified random sampling procedure for our field study. Within the irrigation system, we have covered all districts and distinct agro-climates, and then within the canals, we have covered head reach and tail reach. In order to capture the effect of water control on farming enterprise, gravity irrigation and canal lift were covered. The sampling procedure was discussed in the report. The sample size for each component of the study and for each type of impact, for different locations, is provided in the annexure.

The secondary data on static water levels (SWL) in both the time periods were taken from 222 observation wells. Similarly, the TDS values of water from these 222 observation wells were taken for analyzing the geohydrochemical impacts. The analysis was carried out for seven districts, namely (1) Ahmedabad, (2) Baroda, (3) Banaskantha, (4) Bharuch, (5) Kheda, (6) Mehsana, and (7) Surendranagar.

Study Locations

Overall, the study covered nine districts of Gujarat (Figure 3.2). Of these, seven districts were covered for the survey on the impact of irrigation from the SSP, and two were covered for analyzing the impact of Narmada canal-based water supply project. The seven districts which were covered for the field survey on irrigation impacts are, viz., Bharuch, Vadodara, Narmada, Panchmahals, Ahmedabad, Mehsana, and Surendranagar. In three of these seven districts, that is, in Ahmedabad, Surendranagar, and Mehsana, canal network is not yet ready and farmers are lifting water from the canals directly, and, therefore, its impacts are studied.

Figure 3.2:
Map of Gujarat with Arrows Showing Study Locations

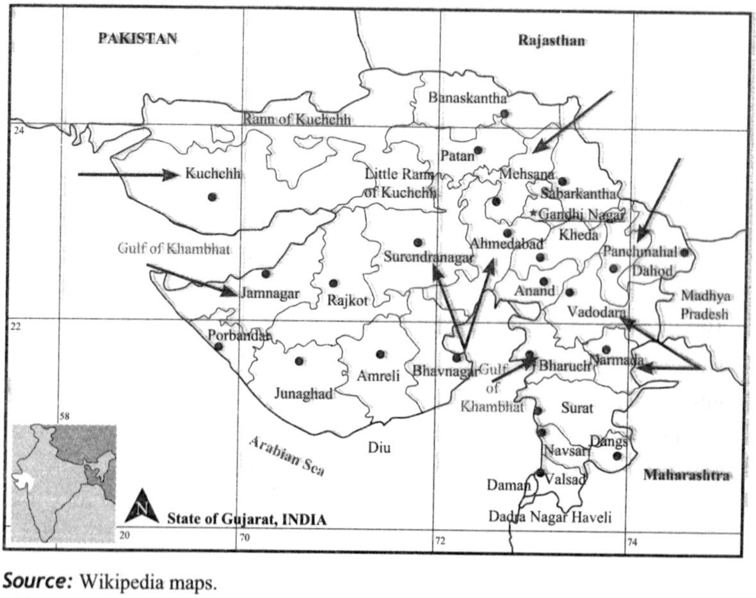

Source: Wikipedia maps.
This map is not to scale and does not depict authentic boundaries.

But, in the districts of Bharuch, Vadodara, Narmada, and Panchmahals, the impact of gravity irrigation through canals was studied. In order to analyze the impact of Narmada canal-based surface irrigation on well irrigators (positive externality of irrigation), surveys were undertaken in six districts, viz., Ahmedabad, Mehsana, Bharuch, Vadodara, Narmada, and Panchmahals. The impact of the Narmada canal-based pipeline water supply scheme was studied in Kutch and Jamnagar districts, including urban and rural areas of Bhuj and Jamnagar.

The location details, including the names of talukas and villages covered for the primary survey required for each type of impact assessment are given in the annexure. The impact of irrigation water supply in the command area on fisheries could not be studied due to the lack of availability of farmers, who were actually engaged in fish production using ponds/tanks. This, however, does not rule out the possibility of such enterprises in the command area.

Caveats of the Study

Much had been discussed in the policy and civil society arena about the negative impacts of the SSP, mainly on the affected and displaced communities and on the environment. A large volume of literature published in the 1980s and 1990s articulated the potential negative socioeconomic impact of the project on the displaced communities, ecological damage due to reservoir submergence and other project interventions, and environmental impacts in the command area due to waterlogging and soil salinity (Baviskar and Singh; 1994; Berger, 1994; Dharmadhikary, 1993; Goyal et al., 1999; Kothari and Bhartari, 1984; Paranjpye, 1990). However, though the SSP has begun to generate numerous socioeconomic and environmental outcomes for the people of the region, this has not received adequate attention from the academic and scientific community. This study is the first attempt to mainly highlight the positive side of the SSP in order to generate a more informed debate. While doing so, it also examines whether the anticipated negative impacts in the command area were really being felt, and how the socioeconomic condition of the displaced population had changed over time. But, it does not attempt to relook at the ecological damage (loss of forest, wildlife and biodiversity) due to reservoir submergence and canal work.

The study involved "longitudinal analysis" using pre- and post-intervention scenarios of all the impact variables for analyzing the changes induced by the SSP, and the effect of changes in external environment during the time period on the impact variables has not been nullified. For instance, it is likely that the improvements in agricultural technologies taking place over time are raising crop yields, though it is the availability of irrigation water which enables farmers to make use of these technologies. These are not factored in our analysis of the impact of irrigation. "Cross sectional analysis" using with and without scenarios of Narmada was also used to revalidate the results obtained from the longitudinal study, though such methods have inherent limitations when used in the context of irrigation enterprise and domestic water supplies. It is extremely difficult to find out identical farmers in terms of farming system characteristics, with one using canal irrigation and the other using the same practices which the present canal irrigator used to follow prior to the introduction of Narmada waters. Again, the soil characteristic and primary productivity of land changed drastically between wetland and the upland outside the design command. Longitudinal analysis used in our study, involving recall method, has its own problems. Nevertheless, past experience undertaking field research on agriculture and water in Gujarat has shown that farmers are able to recall data pertaining to crop inputs and outputs, with reasonable degree of accuracy.

The study is a one-time impact assessment of irrigation and domestic water supplies by comparing the pre- and post-intervention situation at the level of farms and the households of farmers, wage laborers, and domestic water users. Here, both direct socioeconomic impacts and major indirect impacts (positive externalities) of the interventions on well irrigation, energy, drinking water supplies, and carbon emission were also analyzed. Most of the impacts are aggregated for the region under consideration. The study did not intend to capture the changes in the dynamics of farm economy in the neighboring areas/regions due to changes in agricultural practices in the regions studied. For instance, it is possible with hundreds of thousands of farmers in the command area newly taking up irrigation agriculture due to the availability of water;

farmers in the non-command area also get motivated to invest in irrigation systems such as wells and pump sets, and invest in high-yielding crop varieties and crop inputs. What we have followed for analysis is a partial equilibrium model. However, the effect of major social sector interventions like National Rural Employment Guarantee Act has been nullified and these are discussed in the relevant sections.

Annexure: Location Details for Primary Data Collection

Type of Data	Name of District	Names of Talukas	Names of Villages	Sample (Household) Size
Irrigation direct economic benefit: Canal irrigators in SSP command	Bharuch	Bharuch	Dirol, Padariya	180
	Narmada	Tilakvada	Vyadhar	
		Nandod	Vaviyala	
	Panchmahals	Kalol	Sama, Bakrol, Boru, Ratanpur	
	Vadodara	Sankheda	Chhapariya, Alhadapura, Fatepura	
		Bodeli	Vanyadari	
Direct economic benefit from irrigation: Canal lift irrigators in SSP command	Ahmedabad	Dholka	Rupagadh, Ganapatpura, Koth, Kariyana, Balpura, Lana	61
		Bavara	Sakodara	
	Mehsana	Kadi	Maharajpura, Kasava, Narsinhpura, Jayadevpura	61
	Narmada	Tilakvada	Vyadhar	19
		Nandod	Vaviyal	
	Vadodara	Nasavadi	Koyari, Shejakuva, Latipura	56
		Padara		
	Bharuch	Bharuch	Derol, Padariya	21
	Surendranagar	Limdi	Kataria, Tokrala	40
Irrigation—positive externality on well irrigators in SSP command	Ahmedabad	Dholka	Vejalaka, Saradi Rupdadh, Lilapur, Kariyana, Sherapura, Lana, Koth	60
		Bavala	1. Sakodara	

(Annexure Contd)

(Annexure Contd)

Type of Data	Name of District	Names of Talukas	Names of Villages	Sample (Household) Size
	Bharuch	Bharuch	Derol, Padariya, Kissnad, Derol, Parakhet	21
		Aamod	1. Keshalu	
	Mehsana	Kadi	Mahajanpura, Kasava, Sendaradi, Ranchhodpura, Jaydevpura	63
	Narmada	Tilakvada	Vyadhar	58
		Nandol	Vaviyala	
	Panchmahals	Kalol	Ratanpura, Boru, Ramani Muvadi, Sama, Bakrol	29
	Vadodara	Padara	Husepur, Medhad	57
		Sankheda	Vaniyari, Alahadpura	
		Rapar	Medhar	
Indirect benefit from irrigation to wage laborer in SSP command	Ahmedabad	Dholka	Koth, Ganpatpura, Kariyana Kalyanpura	64
		Bavla	Snodra	
	Bharuch	Barucha	Darol	25
		Bharuch	Derol	
	Mehsana	Kadi	Jaydevpura, Thasava, Maharajpura	63
	Narmada	Tilakvada	Vyadhar	60
		Narmada	Vaviyana	
	Panchmahals	Kalol	Sama, Boru	30
	Vadodara	Sankheda	Alhadpura, Variydri, Fatepur Chhatadiya	61
Ecological benefit to the people living on the banks of the river	Ahmedabad/ Gandhinagar	Gandhinagar	Bhat, Koba	20
	Mehsana	Becaraji	Sujanpura	61

(Annexure Contd)

(Annexure Contd)

Type of Data	Name of District	Names of Talukas	Names of Villages	Sample (Household) Size
Socioeconomic impacts & positive	Jamnagar	Jamnagar	Nani Banugar, Moti Banugar, Juna Nagana	59
externalities of drinking water supplies:	Bhuj	Rapar	Chitrod, Badargadh	60
Socioeconomic impacts & positive externalities of drinking water supplies: Urban household	Bhuj	1. Rapar	Alajibapuvas , Uledvas, Zumpadivas, Motovas, Vadivas, Rapar, Ayodhyapuri, Aekatanagar	61
	Jamnagar	Jamnagar	Gulab Nagar, Vibhapara, Moti Banugar	62
Positive externalities of irrigation to domestic water users in SSP command	Vadodara			40

4

Narmada River Basin and Sardar Sarovar Project

Building on its experience in post-independence era, India embarked upon a technologically improved river valley development project, well planned with precisely worked out cost–benefit ratios in Gujarat. Sardar Sarovar is one of the largest multipurpose river valley projects being built with an estimated cost of ₹390 billion (approximately, US $7.8 billion). The project is located on Narmada River, one of the 20 important river basins in India (GOI, 1999). The project was designed to irrigate 1.85 m. ha of land in Gujarat, and supply water for drinking and industrial purposes in the state. About 1,450 MW of hydropower is generated by the project and releases 0.5 MAF of water to Rajasthan. The full irrigation benefits from the project in the state of Gujarat would be derived when water through an elaborate network of primary, secondary, and tertiary canals is released and running to a total of 66,000 km in the state. In order to realize full benefits of hydropower, the height of the dam has to be completed and raised to 138 m.

With some of the most water-scarce regions of the world in its ambit, the project is considered as the lifeline of Gujarat, which is an agriculturally prosperous states in the country. The project had great social, economic, and environmental imperatives for the region. The impoundment of water causes submergence of large area of forest land, and many tribal habitations in Gujarat and MP. Therefore, environmental clearance from the Ministry of Environment and Forests is a necessary requisite for raising the height of the dam. In addition, the construction of the project was opposed through a petition in the Supreme Court by environmental and human right activists on the ground that it would displace indigenous communities living in the catchments. Also, the complex problems associated with land acquisition hamper the progress in completing the canal network.

In this chapter, we would discuss the natural water system of Narmada River basin in which the Sardar Sarovar reservoir is embedded. The discussion is also made on the following: (1) the status of the proposed large water and power systems on the river; (2) the planned utilization of water from Narmada through Sardar Sarovar reservoir; (3) the importance of the project for the social and economic advancement of the state of Gujarat; (4) the complex water distribution and delivery system of the SSP starting from Narmada main canal to branch canals, distributaries to minors, and water courses; and (5) the overall progress in completing the SSP, including its irrigation water distribution and delivery system.

Narmada River Basin

Narmada basin is among the 20 large river basins in India. It lies between 72°32' and 81°45' east longitude, and between 21°20' and 23°45' north latitude. Narmada River, known as the lifeline of MP, is the fifth largest river in India and the largest west flowing river of peninsular part. Against an estimated total annual yield of 1,532 billion cubic meter (BCM) from all the rivers in the subcontinent put together, the net utilizable yield (at 75 percent dependability) of Narmada basin is estimated to be 35.54 BCM or 28 million acre feet (MAF). The Vindhya hills in the north, Satpura ranges in the south, and Maikala ranges in the east form the boundaries of the Narmada basin. The river originates from Maikala ranges at Amarkantak in MP state at an elevation of 900 m. It flows westwards and traverses a distance of 1,312 km before draining into the Gulf of Cambay on the west coast of the country.

The trans-boundary Narmada River basin encompasses four states. The river also forms the boundary of MP and Maharashtra for a distance of 35 km, and Gujarat and Maharashtra for a distance of 39 km. The total drainage area of the basin is approximately 0.1 million sq. km, of which nearly 87 percent is in MP and Chhattisgarh, 11 percent in Gujarat, and 2 percent in Maharashtra. The basin has an elongated shape with a maximum length of 953 km east to west and a maximum width of 234 km north to south. The river has 41 tributaries meeting the trunk river at various confluence points along its course. The shortest is Kharkia River having a length of 25 km and catchment area of 1,099 sq. km, and the longest is Burhner having a length of 177 km and catchment area of 4,228 sq. km.

Water and Power Systems for Narmada Basin

The water and power sector interventions proposed in Narmada basin are vast. As per the Narmada master plan, MP is expected to build a total of 29 major, 135 medium, and 3,000 minor projects in the basin to irrigate 2.75 million hectares of land and generate electricity to the tune of 2,155 MW. The Sardar Sarovar Dam, which is executed in Gujarat, would irrigate 1.8 million hectares of land and generate hydropower to the tune of 1,450 MW per year. Of the 12.03 BCM of water to be harnessed by the SSP, Gujarat is set to supply a total of 11.10 BCM of water. From the remaining, 0.3 BCM go to Maharashtra and the rest 0.6 BCM to Rajasthan. Gujarat's share of the allocation from the SSP has two distinct uses, namely, water for irrigation and water for domestic uses and industry. The total allocation for domestic and industrial purpose is 1.3 BCM. Water for domestic purposes is to be taken to 135 urban centers and 8,215 villages of Gujarat. Nearly 60 percent of these villages are from Saurashtra alone (GOG, 1991).

The extent of water use in the Narmada basin would depend heavily on the progress in a hydraulic asset built in the basin. Of the 29 major projects in MP, only six have so far been completed. Another five are in various stages of construction. This includes Indira Sagar Project, which is very crucial for the SSP, which has experienced huge time and cost overruns. The remaining 18 major projects in MP are yet to be started.

The details of medium and minor irrigation projects are available from the MP Water Resources Department (GoMP, 2001). In the Narmada basin, the irrigation potential created from major schemes is about

0.37 million ha, of which about 0.17 million ha (46 percent) is actually irrigated. The total irrigated area from medium irrigation schemes is 13,000 ha (19 percent) against a created potential of 68,600 ha, while only 24,000 ha (12 percent) is actually irrigated from minor schemes as against a created potential of 0.2 million ha. Therefore, the potential realized is highest for major schemes, followed by medium schemes, and lowest for minor schemes (Table 4.1).

The construction of the SSP dam is nearing completion, though not much is accomplished with respect to the construction of distribution and delivery network for utilizing Gujarat's irrigation share of 9.8 BCM and Rajasthan's share of 0.60 BCM. Gujarat had already built giant pipelines

Table 4.1:
Expected and Current Utilization of Irrigation and Power Potential and Water Resources of the Narmada Basin

Category of Project	No. of Schemes Completed	Area to Be Irrigated (Million Ha)	Area Actually Irrigated (Million ha)	Power Generation Potential (MW)	Power Generation Achieved (MW)	Estimated Water Use (BCM)
29 Major projects	6	1.41	0.37*	2,155.00	90	3.78
135 Medium projects	34	0.67	0.04			0.23
3000 Minor projects	1,252	0.67	0.02			0.11
Total		2.75	0.44	2,155.00	90	4.13
Domestic Industrial Use						
A. Sardar Sarovar Dam	Progress	1.85	0.10	1,450.00		1.21
B. SSP (drinking and industry)	Progress					0.51
Grand Total		4.55	0.54	3,605.00	90.00	5.85

Source: Government of Madhya Pradesh, 2001.
* This is the irrigation potential created; the actual irrigated area data are not available.

to take water from the Sardar Sarovar reservoir to the distant villages in Kutch and Saurashtra for urban and rural drinking water supplies.

In view of the limited progress achieved so far in building water infrastructure, total water utilization through the public irrigation and water supply schemes planned is yet too small (less in comparison to the water utilization potential). We have estimated it to be 5.7 BCM based on three assumptions: (1) the total irrigation water in the basin release to the SSP command is approximately 0.1 million ha out of the total 1.8 million ha planned; delta is close to the design delta; (2) utilization in irrigation against the plan is proportional to the total area irrigated (this closely matches with the actual figures of irrigation water release in the SSP command, reported as 628 MCM during August 2002 and December 2003, if we assume a delta of 0.55 m); and (3) 0.50 BCM out of the plan allocation of 1.3 BCM of water is currently used by Gujarat for water supply to Saurashtra and Kutch. In addition to this, the additional area irrigated in different existing canal commands in central Gujarat through water release from Narmada Main Canal (NMC) is 91,767 ha. The illegal water lifting from NMC for irrigation in central Gujarat is estimated to be 110 MCM per annum. However, this does not include the irrigation use through river lifting on the banks of the river. The total release, therefore, works out to be 1.18 BCM.

The Rationale for the SSP[1]

There are three important facts about Gujarat, which make a strong case for a large water resources project like the SSP. First, there is a large interregional variation in water resources endowment in the state, with one of the highest orographic gradients in the world. While Kutch has a mean annual rainfall of 350 mm, Valsad in south Gujarat receives around 2,000 mm of rainfall on an average (IRMA/UNICEF, 2001; Kumar, 2002). Thus, while south Gujarat is water-abundant, the other three regions are naturally water-scarce. This is compounded by high interannual variability in rainfall and droughts, which are characteristic

[1] This section draws heavily from Kumar and Singh (2001).

of the three regions (Kumar, 2002), significantly upsetting the water balance of these regions. Saurashtra, Kutch, and north Gujarat are highly drought-prone (IRMA/UNICEF, 2001). During drought years, the large surface reservoirs in these regions receive very little inflows and remain more or less dry, and groundwater recharge also suffers. Second, the future demand for water in the state as a whole exceeds the utilizable water resources. More importantly, the regions which are naturally water-scarce have excessively high demands for water, particularly for irrigation. Therefore, the extent of water scarcity is huge in these regions. Third, these regions are already drawing excessive quantities of water from the natural water systems in these regions, causing environmental water scarcity.

The water scarcity problems in Gujarat were analyzed from the point of view of the physical availability of renewable freshwater, available water supplies, and the demand for water from various sectors and the extent of diversion and use of renewable fresh water. The estimates of renewable per capita freshwater availability for the year 2001, water use for the year 1997, and water requirements and supplies for the year 2025 are drawn from the White Paper on Water in Gujarat (IRMA/UNICEF, 2001), details of which are available in the document.

Physical Scarcity of Water

The total renewable freshwater available in Gujarat, including the annual runoff from within Gujarat and that allocated from the neighboring states, and all the natural recharge to groundwater, is 54,593 MCM. This gives a per capita renewable freshwater availability of 904 m^3 per annum for the year 2011. Going by M. Falkenmark's indicator of physical water scarcity, the state is water-scarce (as it is below 1,000 m^3 per capita). But, the availability of water is heavily skewed toward south and central Gujarat which has 69.5 percent of the total renewable freshwater. Therefore, it is important to have regional situation, if one needs to assess the magnitude of water scarcity from a supply-based approach.

Going by regions, the per capita renewable freshwater availability in north Gujarat is only 349 m^3. Thus, the region is "absolutely water-scarce" (as it is below 500 m^3 per capita) where water becomes a constraint to life. Saurashtra and Kutch have per capita renewable freshwater availabilities of 602 m³ and 610m³, respectively. Thus, these regions are "water-scarce" and are approaching the situation of absolute water scarcity (below 1,000 m³ per capita but close to 500m³ per capita) and water becomes a constraint for economic growth and social advancement. South and central Gujarat is "water stressed," with the per capita water availability (1,496 m³) a little below than the threshold of 1,700m³ per annum (see Figure 4.1). Yet, its renewable water availability is four times above the figures for north Gujarat.

However, it is important to note that the estimation based on renewable water availability does not take into account the groundwater stock in regions such as north Gujarat. But, it is quite significant when compared to the renewable groundwater of the region. Therefore, these numbers of annual per capita renewable water availability have to be used very cautiously when one examines the implications of such scenarios for the use of water for economic development in the region. Obviously, the static groundwater resources have played a major role in sustaining the agriculture economy in the region for many decades since its intensive use started. But, today, most of these static resources are mined with the withdrawal of groundwater for agriculture far exceeding the renewable groundwater. Therefore, for projecting future scenarios of human development and economic growth based on water, such numbers would be quite relevant.

Figure 4.1:
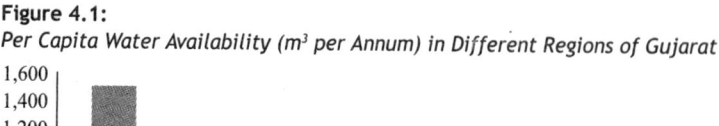
Per Capita Water Availability (m³ per Annum) in Different Regions of Gujarat

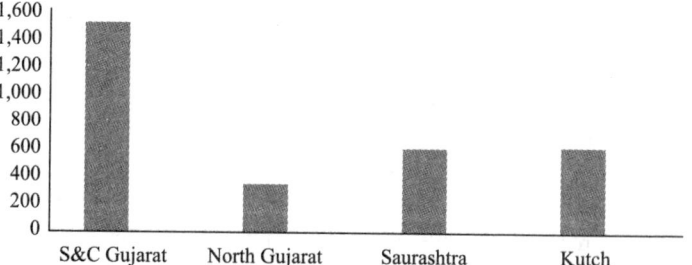

Source: Authors' own estimates based on regional freshwater availability estimates provided in IRMA/UNICEF (2001) and Census 2011.

Demand-driven Water Scarcity

According to the estimates prepared for the White Paper on Water in Gujarat, the total water requirement for agricultural production, domestic and industrial use, and livestock drinking in the year 1997 is 27,622 MCM. The projections showed that by the year 2025, the total requirement of water for agriculture, industry, livestock, and domestic purposes would be 53,050 MCM. Here, it was assumed that the prevailing patterns of water use would continue in the future also, or in other words, a "business as usual" scenario was generated. Under this scenario, we found that even after harnessing all the potentially utilizable surface water and groundwater resources (100 percent of the surface water and 90 percent of the utilizable groundwater) including the water allocated from the neighboring states of MP, Maharashtra, and Rajasthan and water from the SSP (11,100 MCM), the supplies would be only 41,250 MCM. This leaves a gap of 11,837 MCM of water. The situation would appear rather bleak, if one does not consider the allocation of water from Narmada, in which case the magnitude of the gap will be 22,937 MCM.

While a lot of the increase in demand comes from agriculture, a significant portion of the demand is expected to come from urban areas and industries as well. Gujarat today is one of the fastest growing states in the country. The state's urban population is growing rapidly, and the industrial growth is one of the highest in the country. It is also worth noting that many of the large-scale manufacturing units are surfacing in central Gujarat and Saurashtra.

The per capita withdrawal of water is found to be totally unrelated to the relative water availability in a particular region. For instance, the highest withdrawal of water for human uses was found in the case of north Gujarat (448 m³) (IRMA/UNICEF, 2001) which currently has the lowest per capita renewable freshwater (349 m³). The lowest per capita water withdrawal is in south and central Gujarat (372 m³) where the per capita renewable water availability is the highest (1,496 m³). This means that if the current water-use patterns continue, most severe scarcity of water is likely to occur in the region where physical availability of water is limited. The situation can worsen during droughts, when water demand goes up and renewable water resource availability drops drastically leading to acute scarcity situation.

Environmental Water Scarcity

A study commissioned by the United Nation Commission on Sustainable Development (Raskin *et al.,* 1997) had defined water scarcity in terms of the total amount of annual withdrawal as a percentage of the annual water resources. According to this criterion, if the annual withdrawal exceeds 40 percent of the annual water resources, the region can be called water scarce.

Going by this criterion, three regions of the state, namely north Gujarat, Saurashtra, and Kutch are already water scarce and the figures are alarming. In north Gujarat, for instance, the average annual water withdrawal for all uses (estimated to be 6,008 MCM) is already close to the annual water resources (6,105 MCM), throwing clear signals unsustainable resource use. As the runoff (which accounts for 33 percent of the total renewable water) is not fully exploited, the additional water for withdrawal is coming from depletion of the stored groundwater. Alarming drops in groundwater levels manifests this. In Saurashtra, the average annual withdrawal of water is estimated to be 5,449.5 MCM against a total renewable freshwater availability of 9,287 MCM. In Kutch, the average annual withdrawal is 691.7 MCM against an average annual freshwater availability of 1,275 MCM. The withdrawal is exceeding 40 percent of the renewable freshwater in all the three regions, and, therefore, going by the criterion of Raskin *et al.* (1997), these three regions are water scarce.

The Sardar Sarovar Project on Narmada River

The Sardar Sarovar is the last in the series of reservoirs to be built on the Narmada River, and is located in Kevadia village of Narmada district. The gross storage capacity of the reservoir is 9,500 MCM, and the live storage capacity is 5,860 MCM. The total height of the dam from the deepest foundation is 163 m and the full reservoir level is 138.63 m. Sardar Sarovar is a multipurpose project, with planned water release for irrigation, hydropower generation, domestic water supply, and industrial water supply in the state of Gujarat, along with a total allocation of 0.50 MAF of water for Rajasthan through the Narmada main canal.

The NWDT, which came into being in 1969 to adjudicate the dispute over sharing of irrigation and hydropower benefits from the basin between Gujarat and MP, had assessed the dependable yield of the basin as 28 MAF in 1979 for arriving at water allocation decisions for the four party states, viz., Gujarat, MP, Maharashtra, and Rajasthan.

As the final order of NWDT says:

It has been agreed by the party States and decided by the Tribunal in its Order dated 8th October, 1974, that the utilizable quantity of water of 75 per cent dependability in the Narmada at Sardar Sarovar Dam site should be assessed at 28 MAF (34,537.44 MCM).... Out of 28 MAF, 9 MAF has to be provided for Gujarat and 0.5 MAF for Rajasthan at Sardar Sarovar. The requirements at Sardar Sarovar have to be met by releases by Madhya Pradesh and by inflows from the intermediate catchment, surplus to the requirements of Madhya Pradesh below Narmada Sagar and Maharashtra. The releases from Maheshwar work out to 8.12 MAF. Making uniform monthly releases the amount of water to be released by Madhya Pradesh per month would be 834.65 MCM (0.677 MAF).The actual inflow in the river system, however, would vary from year to year and, therefore, the releases by Madhya Pradesh would also vary. (GOI, 1979)[2]

Based on the tribunal award, (1) the 28 MAF utilizable flows of 75 percent dependability in a water year (1st July to 30th June next year) shall be shared by the party states as under: MP—18.25 MAF; Gujarat—9.00 MAF; Rajasthan—0.50 MAF; Maharashtra—0.25 MAF; (2) utilizable surplus or deficit supplies in a water year shall be shared to the extent feasible by the party states in the same proportion as their allotted shares in (1) above. The surplus water shall first be utilized for filling up the reservoirs to capacity and excess water shall be utilized for irrigation and other purposes only after that has been ensured (GOI, 1979).

The project has one main canal which takes water to the farthest points in the command area in the states of Gujarat and Rajasthan. The main

[2] The actual inflow of 75 percent dependability, however, is only 33,316.29 MCM (27.01 MAF) and this is brought up to utilizable quantity of 28 MAF (34,537.44 MCM) by means of carryover in various reservoirs allowing for evaporation losses and regeneration.

canal is a contour canal, with a total length of 458 km. It has a width of 76 m and depth of 7.5 m at the off-take. The NMC alone can store a total of 200 MCM of water, which can ensure drinking water security to the drought-prone areas of the state during the summer month if necessity arises. There are 56 branch canals which off-take from the main canal to feed different agro-climatic zones in all the four regions of the state, including the farthest areas of Kutch. All the branch canals are ridge canals running almost perpendicular to the main canal. The district-wise break up is given in Table 4.2.

Water Distribution System of SSP

The SSP planning involved one of the most testing task of planning and designing water distribution and delivery system for millions of farms spread across the four regions in Gujarat. For the first time in the history of irrigation planning in India, the delivery system till the field inlet was planned in the case of the SSP, and micro-planning was done up to the level of sub-*chak*s of size 5–7 ha within each *chak* (command area of an

Table 4.2:
District-wise Irrigation Benefits from the SSP (Planned)

District Name	Design Command (ha)
Baroda	340,000
Ahmedabad	330,000
Surendranagar	304,000
Banaskantha	313,000
Bharuch	98,000
Mehsana	150,000
Kheda	116,000
Panchmahals	10,000
Gandhinagar	10,000
Bhavnagar	48,000
Rajkot	34,000
Kutch	37,000
Total	1,800,000

Source: GOG (1991).

outlet), having a size of around 40–50 ha.[3] The network planning included deciding the alignment of tertiary canals such as sub-minor/water courses and field channels below the minor outlet. Turnout structures for diverting water from the sub-minor to sub-chaks through field channels were also planned as part of the canal network planning. Irrigation scheduling on the basis of Rotational Water Supply System (RWSS) was also planned in the case of SSP, wherein the timing and duration for which each field under a chak would get water is pre-decided.

It is envisaged that WUAs of beneficiary farmers under each VSA would be formed for managing water distribution. It is expected that the farmers themselves would contribute land and labor for the construction of the field channels and would undertake the responsibility of proper distribution of water.

Hydropower Generation from SSP

The total installed capacity of the hydropower generation system of the SSP is 1,450 MW. It consists of six Francis-type turbines of 200 MW each in the river bed power house (a total of 1,200 MW installed capacity) and five Kaplan-type turbines of 50 MW each in the canal bed power house (a total of 250 MW installed capacity).

Progress in Implementation

The dam construction was done up to a height of 121.92 m by 2006–2007. The construction of the remaining height of the dam is awaiting clearance from the Supreme Court of India on rehabilitation and resettlement of the communities which would be displaced by submergence. The hydropower plants of both river bed power house and canal bed power house are fully functional.

---

[3] Under each village service area (VSA) commanded by a minor, there would be 8–12 chaks each served by a minor outlet with a division box. The VSA becomes part of a block, which a distributary serves.

The construction of the Narmada main canal is already complete and the system has been in operation since 2008. The construction of all branch canals except the Kutch branch canal is also completed, and the progress is around 80 percent. Kutch branch canal is the longest branch canal planned in the entire project and has to pass through one of the ecologically most fragile regions of Gujarat, that is, the little Rann of Kutch. As Table 4.3 shows, out of a total distributary length of 5,112 km, a total length of 2,169 km is already constructed. In the case of minors, only 32 percent of the work is completed, and for sub-minors/watercourses, the progress is even less (21).

With this, the command area in Phase I of the SSP has been receiving water since 2002 with progressive increase in the area commanded over the years. It increased from 1 MAF in 2006–2007 to 3.41 MAF in 2009–2010. Since the Narmada main canal runs through central and north Gujarat before reaching Rajasthan where it serves the command area in that state, hundreds of thousands of farmers in these regions also lift water from the main canal, the branch canals, and distributaries using diesel engines. As a result, the actual extent of irrigation is much higher than the area served by gravity irrigation within designated command area.

***

On its completion, the SSP is expected to add a total of 10,800 MCM of water to Gujarat's water balance. This is nearly 18 percent of the total water demand estimated by the White Paper on Water for Gujarat,

Table 4.3:
Progress in the Construction of Canal Network of the SSP (as on 2012)

Type of Canal	Total Length (km)	Completed Length (km)	Completed (%)
Main canal	458	458	Operational since 2008
Branch canal	2,585	2,050	80
Distributaries	5,112	2,169	42
Minor	18,413	5,821	32
Sub-minor	48,058	10,167	21
Total	74,626	20,685	28

Source: SSNNL, Gandhinagar, 2012.

for the year 2025. Apart from irrigating some of the backward regions of water-rich south Gujarat, it would take water to many of the naturally water-scarce regions which are facing severe shortage of water for irrigation, namely, north Gujarat, Saurashtra, and Kutch. These regions are also drought prone, and during droughts they face acute shortage of water for drinking and domestic uses, and livestock drinking. The communities in these regions mostly depend on underground water resources for domestic water supplies. These reserves have natural contamination with excessively high levels of fluoride and salinity. Thus, they are also highly vulnerable to water-related health hazards. The dependability of flows in Narmada is far higher than that of the rivers in Saurashtra, north Gujarat, and Kutch. As a result, the SSP would be able to release water from the reservoir with a high degree of reliability to these regions and in accordance with water utilization plan. This is expected to reduce the regions' vulnerability to droughts and enhance the livelihoods of millions of farmers who are dependent on crop and livestock production. We would be examining through the next several chapters, whether the project has been able to achieve this goal wherever the water has reached already.

5

Sardar Sarovar Project and Improving Groundwater Regime in Overexploited Regions of Gujarat

Enterprising farmers and rich alluvial aquifers made north and central Gujarat one of the agriculturally most prosperous regions in India. In spite of the low-to-medium rainfalls and high aridity, agriculture in these regions thrived, because of intensive well irrigation. Massive rural electrification, cheap electricity for agricultural sector, good quality power supply, easy access to deep well-drilling technology, and institutional financing were responsible for the development of well irrigation here (Kumar, 2005, 2007). However, there are serious questions raised about the sustainability of agriculture in the region. The groundwater resources in these regions, which are the backbone of the region's agriculture and dairy production, are fast depleting, as evidenced by the rapidly falling water levels and deteriorating water quality with increasing levels of TDS (IRMA/UNICEF, 2001) until the introduction of Narmada waters.

With limited water available from the surface irrigation schemes, farmers in the region depend excessively on groundwater resources to irrigate agricultural crops and fodder. The cheap electricity by the State Electricity Board supplied on flat rate basis for agriculture provided great impetus for farmers to pump groundwater not only to irrigate their own crops, but to sell it to the neighbors, leaving no incentive to use it efficiently. These tube wells account for about 3,000 MCM of annual gross groundwater draft resulting in an annual groundwater deficit of about 600 MCM. The direct economic costs of groundwater depletion had been of great concern to the state government which subsidizes electricity use in north Gujarat's tube well irrigation to a tune of ₹3 billion per year. For farmers too, agriculture is becoming increasingly unprofitable because of the high costs of groundwater irrigation which, in spite of highly subsidized electricity, are so high that the cultivation of all crops—barring a few needing sparse irrigation—is already unviable in this region (IRMA/UNICEF, 2001).

In the past, numerous ideas of water imports in Gujarat have been floated, because of the uneven distribution of water resources between south Gujarat and the rest of the state. The state has a wide range of hydrological conditions with rainfall varying between 300 mm in Kutch to 2,000 mm in the Dangs. The grandiose Kalpasar project was conceived to store surplus waters of the rivers to the north of Narmada in a gigantic fresh water reservoir by building a dam in the Gulf of Cambay and to use the water to fulfill the needs of Saurashtra. Closer to the water import exercise , there have been separate proposals by Vitthalbhai Patel, a noted Gujarati thinker, and Khemabhai Patel, a farmer leader of Sabarkantha, to link the hydrologic systems of north and south Gujarat to reduce water stress in the former (Shah, 1998). The Sardar Sarovar Development Plan itself had envisaged the use of water pumped from downstream of the main dam after use in power generation and supplied to three major and nine medium projects in north Gujarat, to augment their water storage capacity and, thus, provide security to about 2.3 lakh ha irrigated land under these schemes (GOG, 1989). A study carried out by Tahal Consulting Engineers for the Government of Gujarat looked specifically at the topic of artificial recharge in north Gujarat and recommended the use of imported water from Narmada for recharge through "spreading" in upper regional aquifers and river beds of rivers (GOG, 1996).

With the Sardar Sarovar Dam reaching a stage of completion where it is possible to divert water into the main canal, water import now is a reality. With inflows to the dam being expected to exceed Gujarat's allocation of 9.0 MAF during all better-than-average monsoon years, tremendous scope exists for using this water for ecological purposes, such as recharging north Gujarat's groundwater. These measures, if carried out systematically, can have long-term positive impacts from the point of view of social, economic, and environmental changes. Therefore, it is important to assess the geo-hydrological and water quality impacts of current water imports in alluvial north and central Gujarat, and to qualitatively assess the welfare effects these changes induced in the irrigation and the drinking water economy by this water import.

Impact of Surface Irrigation from SSP on Groundwater Regime in Gujarat

There are several ways by which the SSP can influence on the groundwater regime of Gujarat.

First, water released through Narmada main canal is being lifted by farmers in the area close to both banks of the main canal for irrigation (Kumar and Bassi, 2011). In the process, they would eventually replace groundwater-based sources for irrigation with canal water. Because of this, there would be reduced groundwater pumping in the area benefited by canal lifting. This benefit would be available in at least 900–1,800 sq. km of central and north Gujarat. The fact that the main canal has a length of 532 km and runs across the most water-scarce regions of the state when considered, the recharge would benefit the unconfined alluvial aquifers of central and north Gujarat, given the alignment of the main canal, while reduced groundwater abstraction would greatly benefit the confined aquifers underlying the phreatic (unconfined) aquifer.

Second, water currently being released into several rivers of central and north Gujarat, from the left bank of the main canal, acts as an excellent source of recharge for the alluvial aquifers of the Cambay basin, by virtue of the thick sand bed overlaying these rivers. The area, which is affected by the discharge into the rivers, would be quite substantial, and there are many such rivers in central and north Gujarat, flowing in the northeast–southweat (NE–SW) direction, and the formation of the entire region is (unconsolidated) alluvium with sandy and sandy loam soils which allow fast recharge of percolating water. The river course is perpendicular to the main canal alignment. This again would benefit the unconfined aquifers, which in term can increase the leakage to the lower confined aquifers.

Third, water which is being released in the first phase of the SSP command for the past 7–8 years, in Vadodara, Narmada, and Bharuch districts, and in part of the second phase in the recent years, mostly in Ahmedabad and Bhavnagar districts, would recharge groundwater through return flows from the irrigated fields and seepage from the canals supplying irrigation water, while there would be a reduction in groundwater draft for irrigation in these areas. The first phase command covers an area of 0.4 million ha. Hence, the area benefited would be quite substantial.

There are many prerequisites for groundwater recharge to be physically, economically, and environmentally sustainable. The first condition is that the soils have to be highly permeable to allow continuous infiltration of the impounded water and the vertical hydraulic conductivity is good to allow deep percolation of the infiltrating water. The vertical hydraulic conductivity of the phreatic aquifer in north Gujarat is estimated to be 0.0003 m per day (Kavalanekar *et al.*, 1992).

The second condition is that the aquifers have sufficient empty storage space. As regards the first one, the soils in north and central Gujarat have high permeability and infiltration capacity. The porosity is estimated to be in the range of 0.10 in the upper strata to 0.03 in the lower strata (Kavalanekar *et al.*, 1992). As regards the second condition, Ranade and Kumar (2004) pointed the favorable geo-hydrological condition in north Gujarat for sustained groundwater recharge. Similar situation exists in central Gujarat also, where a trough is formed in the alluvial aquifer around Ahmedabad due to the mining of groundwater (CGWB, 1998). The third factor is the vertical distance between the crop root zone and the saturated zone or the level to which water has to move vertically down or the thickness of the unsaturated zone. In the case of north and central Gujarat, the upper aquifer, which has a depth of around 100 feet, is separated from the lower confined aquifers by a semi-pervious layer. Hence, return flow benefits would be significant.

The last factor is the degree of saturation of the unsaturated zone. Going by Richards' equation, which governs the groundwater flow dynamics in unsaturated zones, higher the degree of moisture content or saturation of the unsaturated zone, higher would be the hydraulic conductivity for the same level of permeability (Watt, 2008). Hence, the speed of the movement of return flow water through the vadose zone depends on the time at which it takes place. If it occurs during the rainy season, there would be faster movement of the return flow. It is important to note here that a large proportion of the water delivered from the Sardar Sarovar reservoir through canals into the rivers and command area is during monsoon season, increasing the rate of return flow.

Additional recharge to groundwater from the relatively better quality water from Narmada River, particularly from the point of view of chemical properties, also shows that there would be greater dilution of natural groundwater in these regions, with changes in mineralogical composition of the water in the aquifers. The aquifers in north and central Gujarat have high levels of fluoride and TDS, as expected.

Geo-hydrological Impacts of Introduction of Narmada Waters

The estimates of change in groundwater levels in the command area of the SSP covering six districts of Gujarat, during the two time periods for both pre- and post-Narmada water introduction, are presented in Tables 5.1, 5.2, and 5.3. Table 5.1 present outputs of analysis for the two time periods January 1996–January 1999 and January 2005–January 2010.

Table 5.1:
Change in the SWL (January)

District	January 1996–1999 (Pre-Narmada Water Introduction)		January 2005–2010 (Post-Narmada Water Introduction)	
	Average Rise/ Fall in SWL (m)	*Average Rate of Rise/Fall in SWL*	*Average Rise/ Fall in SWL (m)*	*Average Rate of Rise/Fall in SWL*
Ahmedabad	2.212	0.737	1.197	0.239
Banaskantha	−5.747	−1.916	0.458	0.092
Baroda	1.814	0.605	6.177	1.235
Bharuch	0.850	0.283	2.205	0.441
Kheda	13.066	4.355	7.628	1.526
Mehsana	2.131	0.710	11.200	2.240
Surendranagar	2.125	0.708	4.664	0.93

Source: Authors' own analysis based on data shared by SSNNL.

Table 5.2:
Change in the SWL (May)

District	May 1996–1998 (Pre-Narmada Water Introduction)		May 2004–2009 (Post-Narmada Water Introduction)	
	Average Rise/ Fall in SWL (m)	*Average Rate of Rise/Fall in SWL*	*Average Rise/ Fall in SWL (m)*	*Average Rate of Rise/Fall in SWL*
Ahmedabad	3.050	1.525	4.066	0.813
Banaskantha	–1.242	–0.621	0.764	0.153
Baroda	2.827	1.414	6.172	1.234
Bharuch	1.800	0.900	4.294	0.859
Kheda	6.736	3.368	5.694	1.139
Mehsana	3.178	1.589	10.946	2.189
Surendranagar	2.292	0.917	5.943	1.189

Source: Authors' own analysis based on data shared by SSNNL.

Table 5.3:
Change in the SWL (October)

District	October 1996–1998 (Pre-Narmada Water Introduction)		October 2004–2009 (Post-Narmada Water Introduction)	
	Average Rise/ Fall in SWL (m)	*Average Rate of Rise/Fall in SWL*	*Average Rise/ Fall in SWL (m)*	*Average Rate of Rise/Fall in SWL*
Ahmedabad	0.871	0.435	4.557	0.911
Banaskantha	–0.596	–0.298	0.236	0.047
Baroda	2.090	1.045	7.675	1.535
Bharuch	1.167	0.584	5.742	1.148
Kheda	9.689	4.845	7.261	1.452
Mehsana	4.843	2.651	7.637	1.527
Surendranagar	2.322	1.161	8.197	1.639

Source: Authors' own analysis based on data shared by SSNNL.

Table 5.2 presents the outputs of analysis for May 1996–1999 and May 2004–May 2009. Table 5.3 presents the analysis for October 1996–October 1999, and October 2004–October 2009. Both the aggregate rise/fall and annual rate of rise/fall are presented in the Tables.

In Ahmedabad district, the average rate of rise in SWL was observed to be 0.74 m in the pre-Narmada water scenario, but came down to 0.24 m in the post-Narmada scenario (Table 5.1). The average rate of rise in Banaskantha district went up from –1.92 m in the pre-Narmada to 0.092 m in the post-Narmada scenario. Baroda district recorded an average rate of rise in SWL of 0.605 m prior to the introduction of Narmada waters and 1.235 m in post-Narmada water introduction, showing a faster rise in water levels. In Bharuch district, prior to the introduction of Narmada waters, the average rate of rise in water levels was observed to be 0.28 m, but went up to 0.44 m in post-Narmada water introduction.

The average rate of rise in SWL at Kheda district was observed to be 4.35 m prior to Narmada water introduction and came down to 1.53 m post-Narmada water introduction. Mehsana district recorded an average rate of rise in SWL of 0.71 m prior to Narmada water introduction, but the rate of rise increased to 2.24 m post-Narmada water introduction. Prior to the introduction of Narmada waters, the average rate of rise in SWLs in Surendranagar district was observed to be 0.71 m, which increased to 0.93 m in post-Narmada water introduction.

Overall, the water level trend of January is positive, with the depth to water levels reducing. More importantly, the annual rate of rise in water levels increased post-Narmada in Baroda, Bharuch, Mehsana, and Surendranagar, whereas there is a positive change (reversal) in the water level trend in Banaskantha. In Ahmedabad district, the average annual rate of rise in water levels declined post-Narmada. One district where the rate of rise in water levels had declined substantially is Kheda.

From May to May in a year, the average rate of rise in the SWL in Ahmedabad district was observed to be 1.53 m prior to Narmada water introduction, but reduced to 0.81 m post-Narmada (Table 5.2). In Banaskantha district, the average rate of rise in the SWLs increased from –0.62 m (fall) prior to the introduction of Narmada waters to 0.15 m after Narmada water introduction, showing a trend reversal. Baroda district had an average rate of rise in the SWL of 1.41 m per year prior to Narmada waters and came down slightly to 1.23 m post-Narmada water introduction. In Bharuch district, the average rate of rise in the SWLs went down slightly from 0.90 m prior Narmada water introduction to 0.86 m post-Narmada water introduction.

Kheda district measured an average rate of rise in the SWL of 3.37 m in pre-Narmada water scenario and 1.14 m after the introduction of Narmada waters. The average rate of rise in the SWL at Mehsana district is from 1.59 m prior to Narmada waters introduction to 2.19 m post-Narmada water introduction. Similarly, in Surendranagar district, the average rate of rise in the SWL is from 0.92 m to 1.19 m post-Narmada water introduction, respectively.

Overall, the water level trend for the month of May is positive, with the depth to water levels reducing, though the rate of rise in water levels was a little lower post-Narmada in four districts. The difference is not very significant in Baroda and Bharuch. The sharp decline in the rate of water level rise found in the case of Kheda district can be attributed to the zero impact of Narmada. A large part of the district receives water from the Mahi canal system and any changes in the water release from Mahi can have serious impacts on groundwater in the area, as reduced canal water would motivate farmers to increase their dependence on groundwater. Importantly, the average rate of rise in water levels had increased in Mehsana and Surendranagar districts and there is a positive reversal of water level trend in Banaskantha.

As Table 5.3 shows, Ahmedabad recorded an average rate of rise in the SWL from 0.44 m to 0.91 m post-Narmada water introduction. The average rate of rise in water levels in Banaskantha district increased from –0.29 m to 0.047 m post-Narmada period. Baroda district recorded an average rate of rise in the SWL from 1.05 m prior to the introduction of Narmada waters to 1.54 m.

The average rate of rise in the SWL in Bharuch district increased from 0.58 m to 1.15 m post Narmada water introduction, whereas in Kheda district the average rate of rise in the SWL decreased from 4.845 m to 1.452 m. In Mehsana district also the average rate of rise in the SWL decreased from 2.65 m to 1.53 m. Surendranagar district recorded an average rate of rise in the SWL from 1.16 m to 1.64 m.

Overall, the water level trend of October is positive, with the depth to water levels reducing. More importantly, the rate of rise in the water levels had increased post-Narmada water introduction in Ahmedabad, Baroda, Bharuch, and Surendranagar districts, whereas there is a positive change in the water level trend in Banaskantha. The only district where the rate of rise in water levels had reduced, that too substantially, is Kheda district.

Figures 5.1, 5.2, and 5.3 represent a change in the SWLs prior to Narmada water and post-Narmada waters, covering three seasons

Figure 5.1:

Average Rise/Fall in SWL (January)

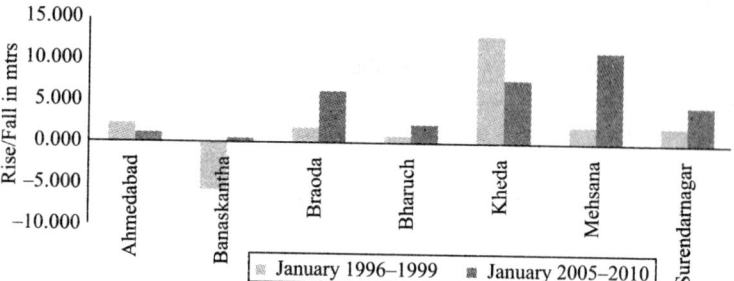

Source: Author.

Figure 5.2:

Average Rise/Fall in SWL (May)

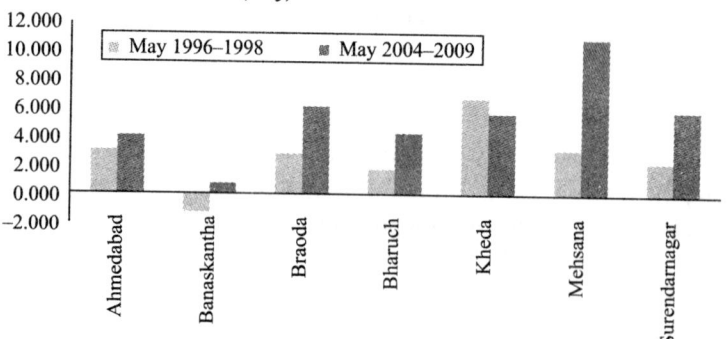

Source: Author.

Figure 5.3:

Average Rise/Fall in SWL (October)

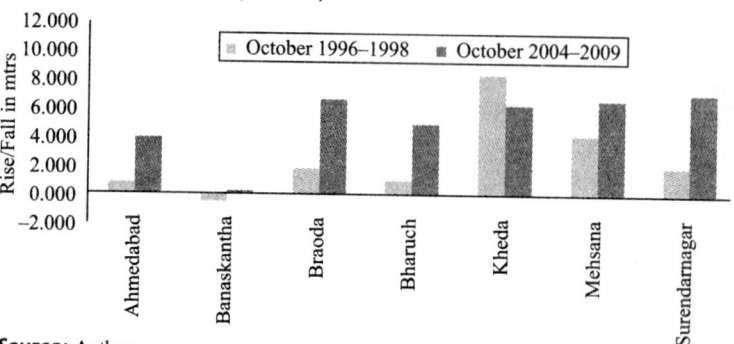

Source: Author.

(winter, pre-monsoon, and post-monsoon). The negative values indicate fall in water table and positive values indicate rise.

It can be seen from Figures 5.1, 5.2, and 5.3 that post-Narmada the overall trend in water levels in all the districts except Kheda is not only upwards but also higher as compared to the pre-Narmada period.

Environmental Impacts on the Hydrogeology of Groundwater

The geo-hydro chemical impacts are analyzed by comparing the average rate of increase/drop in TDS values in both the periods of Narmada waters. The average rise/drop in the TDS value for three seasons (January, May, and October), in seven districts for both the scenarios are presented in Tables 5.4, 5.5, and 5.6.

TDS (January)

As Table 5.4 shows, Ahmedabad district recorded an average rate of the TDS value of 283 ppm prior to Narmada water introduction. But, the TDS recorded a fall of 32.6 ppm per year during the post-Narmada water introduction. The average rate of change in the TDS value in Banaskantha district decreased from 71.7 ppm per year (rise) prior to Narmada water introduction to 285.1 ppm (fall) per year post-Narmada

Table 5.4:
TDS (January)

District	January 1996–1999 (Pre-Narmada Water Introduction)		January 2005–2010 (Post-Narmada Water Introduction)	
	Average Rise/ Fall (ppm)	*Average Rate of Rise/Fall*	*Average Rise/ Fall (ppm)*	*Average Rate of Rise/Fall*
Ahmedabad	−849.231	−283.1	163.0	32.61
Banaskantha	−215.000	−71.7	1425.7	285.1
Baroda	530.000	176.7	648.4	129.7
Bharuch	3507.857	1169.3	2065.0	413.0
Kheda	−122.222	−40.7	90.0	18.0
Mehsana	3520.000	1173.3	−300.0	−60.0
Surendranagar	5868.333	1956.1	675.7	135.14

Source: Authors' own analysis based on data shared by SSNNL.

Table 5.5:
TDS (May)

District	May 1996–1998 (Pre-Narmada Water Introduction)		May 2004–2009 (Post-Narmada Water Introduction)	
	Average Rise/ Fall (ppm)	Average Rate of Rise/Fall	Average Rise/ Fall (ppm)	Average Rate of Rise/Fall
Ahmedabad	84.231	42.115	9,735.308	1,947.062
Banaskantha	–61.311	–30.656	1,765.000	353.000
Baroda	1,560.357	780.179	1,024.828	204.966
Bharuch	3,476.333	1,738.167	2,296.667	459.333
Kheda	–55.000	–27.500	–7.500	–1.500
Mehsana	–178.750	–89.375	–190.000	–38.000
Surendranagar	5,086.667	2,543.333	–1,933.000	–386.600

Source: Authors' own analysis based on data shared by SSNNL.

Table 5.6:
TDS (October)

District	October 1996–1998 (Pre-Narmada Water Introduction)		October 2004–2009 (Post-Water Introduction)	
	Average Rise/ Fall (ppm)	Average Rate of Rise/Fall	Average Rise/ Fall (ppm)	Average Rate of Rise/Fall
Ahmedabad	208.183	104.092	647.037	129.407
Banaskantha	164.513	82.256	–1,544.286	–308.857
Baroda	–761.000	–380.500	514.225	102.845
Bharuch	95.172	47.586	78.182	15.636
Kheda	–213.000	–106.500	5.000	1.000
Mehsana	3,100.000	1,550.000	450.000	90.000
Surendranagar	4,900.000	2,450.000	496.842	99.368

Source: Authors' own analysis based on data shared by SSNNL.

water introduction. Baroda district recorded an average rate of fall in TDS of 176.7 ppm, to 139.7 ppm post Narmada water introduction. The average rate of fall in the TDS values in Bharuch district declined from 1,169.3 ppm (prior to Narmada water) to 413 ppm post-Narmada.

In Kheda district, the rate of rise in the TDS value was observed to be 40.7 ppm per year prior to Narmada water introduction, but started declining at a rate of 18 ppm post-Narmada. The average rate of change in

TDS in Mehsana district decreased from 1,173 ppm per year to –60 ppm per year post-Narmada. Similarly, Surendranagar district also recorded an average rate of drop in TDS of 1,956 ppm, to 135.1 ppm with the advent of Narmada waters.

Overall, in six out of the seven districts, the TDS values either continued to decline at lower rates (Bharuch, Baroda, and Surendranagar) or started showing positive reversal of trend (Ahmedabad, Kheda, and Banaskantha) post-Narmada. The only district, where the change in trend was negative, is Mehsana.

TDS (May)

As Table 5.5 shows, the average rate of fall in TDS in Ahmedabad district increased from 42 ppm prior to Narmada water introduction to 1,947 ppm post-Narmada. Banaskantha district recorded a trend reversal, with the TDS dropping at a rate of 353 ppm post-Narmada water introduction, while the TDS increased at an annual rate of 30.6 ppm prior to Narmada water introduction. The average rate of fall in TDS in Baroda district was observed to be 780.1 ppm before the introduction of Narmada waters, which decreased to 204.9 ppm post-Narmada water introduction. Bharuch district recorded a decline in the average rate of fall in TDS with an annual rate of drop of 1,748.1 ppm prior to Narmada water introduction and 459.3 ppm post-Narmada water introduction.

In Kheda district, the average rate of increase in TDS declined from 27.5 ppm to 1.5 ppm in post-Narmada times. In Mehsana district, a similar trend from 89.4 ppm to 38 ppm was witnessed. Surendranagar district recorded an average rate of fall in TDS of 2,543.3 ppm prior to Narmada water introduction, but the trend got reversed with the TDS increasing at a rate of 386 ppm post-Narmada.

Overall, post-Narmada, the TDS values of groundwater in the designated command area either continued to decline at lower rates (Baroda and Bharuch) or higher rates (Ahmedabad), or started showing positive reversal of trend (Banaskantha) or the rate of increase in TDS got lowered (Kheda and Mehsana). Thus, in six out of the seven districts, the trend was positive. The only exception was Surendranagar district where the trend got reversed from drop in TDS during the pre-Narmada period to rise in TDS post-Narmada.

TDS (October)

Table 5.6 shows that Ahmedabad district recorded an average rate of fall in TDS of 104.1 ppm prior to the introduction of Narmada waters, which went up slightly to 129.4 ppm post-Narmada. The average rate of fall in TDS at Banaskantha district was observed to be 82.2 ppm before the introduction of Narmada waters, but the trend got reversed with the annual rate of increase in TDS touching 308.8 ppm post-Narmada. Baroda district recorded an average rate of rise in TDS of 380 ppm prior to Narmada water introduction, but the trend got reversed with the TDS values reducing at a rate of 102.8 ppm post-Narmada water introduction.

In Bharuch district, the average rate of drop in TDS values decreased from 47.5 ppm prior to Narmada water introduction to 15.6 ppm post-Narmada. In Kheda district, while the TDS values increased during the pre-Narmada period at an average rate of 156 ppm per year, it started declining at a marginal rate of 1.0 ppm post-Narmada water introduction. In Mehsana district, the average rate of drop in TDS declined from 1,550 ppm per year prior to Narmada water introduction to 90.0 ppm per year post-Narmada. The average rate of fall in TDS in Surendranagar district was observed to be 2,450 ppm per year prior Narmada water introduction, but went down to 99.4 ppm per year post-Narmada water introduction.

The overall trend is that in six out of the seven locations, post-Narmada, the average TDS values of groundwater either continued to decline, at lower rate in some cases (Bharuch, Mehsana and Surendra-nagar) and higher rates in some others (Ahmedabad), or started showing positive reversal of trend (Baroda and Kheda). The only exception was Banaskantha district, where the TDS values rose by nearly 1,544 ppm over a period of five years post-Narmada water introduction.

The average rate of increase/decrease in TDS values for three seasons, viz., January, May, and October in the seven districts are shown in Figures 5.4–5.6. Here, the positive values indicate reduction in TDS and the negative values indicate increase in TDS.

Figure 5.4:
Average Rise/Fall in TDS Value (January)

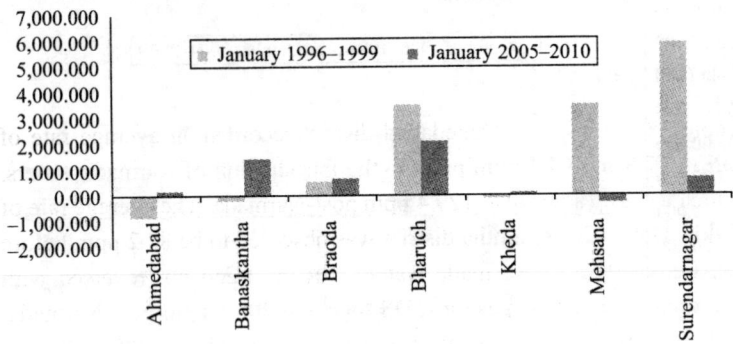

Source: Author.

Figure 5.5:
Average Rise/Fall in TDS Value (May)

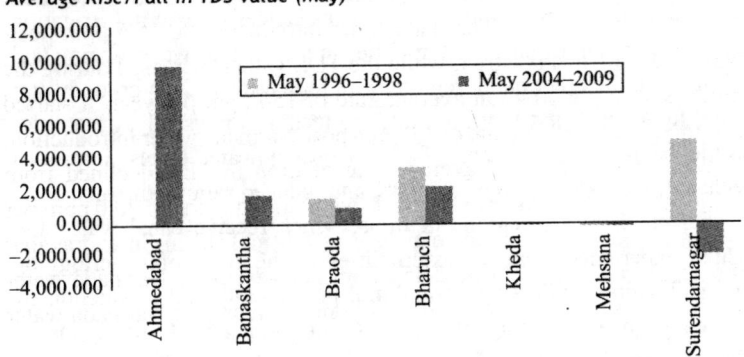

Source: Author.

Potential Welfare Effects of Surface Water Import

Introduction of surface water in an area need not always produce positive externalities on the society. The externality induced by canal irrigation can be negative as well. The physical or technical externalities, which translate into economic externalities, are governed by rainfall, climate, soil types—whether heavy soils or light soils—and geo-hydrology of the area—depth to groundwater table, quality of groundwater, etc.

Figure 5.6:
Average Rise/Fall in TDS Value (October)

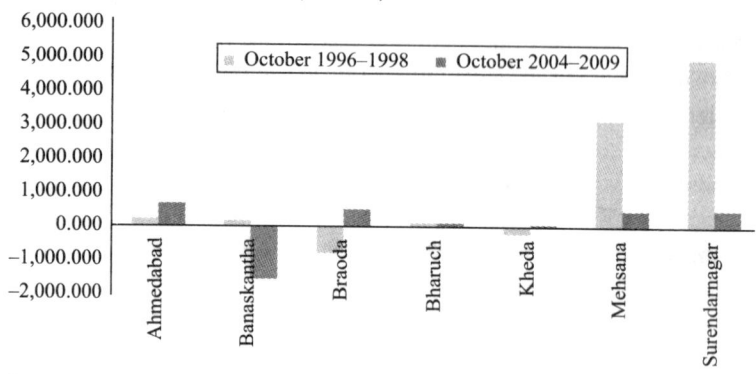

Source: Author.

Obviously, in an area which has heavy soils, shallow groundwater table, and high rainfall, there are greater chances of external irrigation inputs generating negative externalities in the form of waterlogging and salinity. On the other hand, regions with high aridity, deep groundwater table, light soils, and low rainfalls, the technical/physical externalities would be positive in terms of rise in groundwater levels, improved recharge, increased yield of aquifers, and reduced salinity of soils.

In fact, there are regions in India, which receive canal water and which experience negative externalities in the form of waterlogging and soil salinity, affecting agricultural productivity and production by reducing the productivity of the soils and making land totally unsuitable for crop production. Two examples to cite here are Sone command in Bihar (Meinzen-Dick, 1995) and Mahi command in south Gujarat (IRMA/UNICEF, 2001). On the other hand, there are regions which are benefited by the introduction of canal water, in the form of improved sustainability of well irrigation. For example, the Mulla command in Maharashtra, the command area of Krishna delta in Andhra Pradesh, and the command areas in Punjab and Haryana. Without imported surface water, which provide additional recharge in the area (Kumar *et al.*, 2010a), groundwater use in these regions would have become unsustainable much too early.

However, it does not take much effort to draw the inference that the likely impacts of the introduction of canal water in north and central

Gujarat would be positive. Following are the reasons. (1) The aquifers are heavily depleted in the region (Kumar, 2007; Ranade and Kumar, 2004), which entirely rules out the chances of waterlogging and salinity. (2) The region is "absolutely water scarce," with per capita renewable water availability below the 500 m^3 per annum mark in north Gujarat, whereas the region from where water is exported is "water rich" with renewable water availability of 1,496 m^3 per annum, as presented in Chapter 2. (3) The region experiences high degree of water scarcity from the point of view of demand and supplies (Kumar and Singh, 2001; Kumar, 2002), meaning that the use the value of water in the region would be far higher than that in south Gujarat (Kumar et al., 2008). (4) The region experiences environmental water stress, meaning that the availability of water in the rivers, etc., would provide great environmental and ecological benefits (Kumar and Bassi, 2011).

Theoretically, augmentation of groundwater through irrigation return flows in the region would benefit the society economically in terms of reduction in the cost of pumping groundwater for irrigation, increase in yield and, therefore, irrigation potential of wells and the overall economic surplus from well irrigation, and reduction in the incidence of well failures. This happens in practice also, as both the command area and area irrigated through canal lifting have irrigation wells, which are facing problems of reducing yield and failures due to groundwater overdraft.

Over and above, the quality of groundwater in central and north Gujarat is not good, with incidence of high levels of TDS in most parts and fluoride in certain pockets (IRMA/UNICEF, 2001). Increased recharge can cause significant dilution of the mineral containing groundwater. This can render it suitable for drinking. Currently, many villages and towns in the region are supplied drinking water from distant sources, including the SSP. Therefore, the impact of quality amelioration would be in the form of reduction in investment for the provision of drinking water supplies in villages and urban areas of the region. This is in addition to the benefit of the reduced cost of pumping water for domestic water supplies in rural and urban areas of the regions which are likely to be benefited by canal irrigation, and the reduced cost of the import of drinking water through tankers during summer months because of groundwater scarcity.

Findings

To sum up, there has been a significant difference in groundwater behavior in the designated command areas of the SSP between the two time periods, that is, pre-Narmada and post-Narmada. The districts covered in the analysis are Banaskantha, Mehsana, Ahmedabad, Surendranagar, Baroda, Bharuch, and Kheda.

Season-wise analysis shows that everywhere except Kheda, where the water levels showed remarkable rising trends during the pre-Narmada period, the water level either started rising at a faster rate or the rising trend continued at a slightly lower rate post-Narmada, or the trend got reversed from the lowering trend (pre-Narmada) to the rising trend (post-Narmada).

Out of these three different patterns, the first and the third can be treated as the positive impacts of the introduction of exogenous water from the SSP. As regards the second, it was encountered in Ahmedabad during January 2005–January 2010 and May 2004–May 2009, in Mehsana during October 2004–October 2009, and in Baroda and Bharuch during May 2004–May 2009. If we consider the fact that the demand for groundwater is on the rise with the passing of time with a resultant increase in groundwater draft, the slight decline in the annual rate of rise in water levels is quite possible even with additional recharge from Narmada waters.

Whereas in the case of Kheda, the pre-Narmada trend in water levels showed significant rise at a rate of 4.34 m per year, which dropped significantly post-Narmada to 1.53 m per year, hence, Narmada does not seem to have any impact on groundwater in the region.

Interestingly, the changes in groundwater level trends during the post-Narmada water period in most districts were not consistent across seasons. The only two districts where the trend in the water level was consistent were Surendranagar and Banaskantha. In all other districts, the trend was inconsistent. For instance, in Ahmedabad district, the rate of rise in water levels was lower during May 2004–May 2009 and January 2005–January 2010, as compared to the periods, May 1996–May 1999 and January 1996–January 1999. But, the rate of rise in water levels was higher during October 2004–2009 as compared to the corresponding period, that is, October 1996–October 1999. Similar inconsistency was seen in the case of Baroda, Bharuch, and Mehsana. Such an inconsistent trend could be due to sharp seasonal variations in water levels within the same year.

The areas receiving Narmada waters either through canals or through river discharge appear to have actually been benefited in terms of improved groundwater recharge, with the exception of Kheda. In Kheda district, while the decline in the rate of rise in water levels is significant, this cannot be inferred as an impact of Narmada waters. The fact is that there are factors other than Narmada which actually influence the groundwater level trends in Kheda district,[4] while area commanded by Narmada is only a small fraction of the district area.

As regards groundwater quality, the overall water quality trend during the period of post-Narmada water introduction is positive, with the average TDS values declining consistently over a period of time in most districts or the rate of increase in TDS values declining. Comparing this trend with the pre-Narmada trend shows the following outcomes. In districts where the TDS had reduced prior to Narmada, the "lowering trend" continued post-Narmada also, but with higher or lower annual rate of drop. The first type of situation was encountered in Ahmedabad during January 2005–January 2010 and May 2005–May 2010 as compared to January 1996–January 1999 and May 1996–May 1998. The second type of situation was encountered in Bharuch district irrespective of the season considered for analysis. In Baroda district, the same situation was encountered during January 2005–January 2010 and May 2004–May 2009 as against January 1996–January 1999 and May 1996–May 1999, respectively.

In two districts, viz., Kheda and Ahmedabad, there has been a trend reversal, with the rising trend being replaced by a sliding trend. Only in one district, that is, Surendranagar, the trend reversal was negative. But, what is most interesting is that the trend in most districts is not consistent across seasons like what was found in the case of water level trends.

---

[4] The most important being the annual variations in the volumetric release of water from Kadana reservoir for the Mahi irrigation scheme, which cover a large area in Kheda district, and the consequent changes in groundwater abstraction by farmers in the command.

The two districts where the "positive trend" of higher rate of reduction in TDS values or annual increase in TDS getting replaced by annual reduction in TDS was consistent across seasons are Ahmedabad and Kheda. Bharuch is the only district where the negative trend of lower rate of reduction in TDS was found to be consistent irrespective of the season chosen for analysis.

There would be two major indirect impacts of the introduction of canal water. The first type of impact would be on the irrigation sector, resulting from improved sustainability of well irrigation in the form of increase in the irrigation potential of wells, reduction in the incidence of well failures, and the reduced cost of energy for pumping groundwater for irrigation. The second type of impact will be on drinking water economy, resulting from: (1) the improved quality of groundwater, which reduces the investments required for treating it for making it potable for local tube well-based water supplies, and (2) the rise in water table and improved replenishment of aquifers, reducing the cost of electricity required for pumping groundwater for rural and urban water supplies, and deferring the investment required for transport of tanker water during summer months due to drying up of local drinking water sources.

***

The decrease in average rate of drop in the SWLs after the introduction of Narmada can be attributed to the improved water levels in the monitoring wells' waters owing to irrigation return flows from canal commands and seepage from canals or reduced groundwater pumping for irrigation or both. The decrease in the average rate of rise in the TDS levels can be attributed to improvement in the quality of groundwater due to dilution, owing to return flows from the irrigated fields, or reduced salinization of pumped aquifers resulting from reduced withdrawal, or both. The rise in water levels can bring about the following economic impacts: (1) increased availability of renewable groundwater across seasons, (2) reduced cost of pumping groundwater, and (3) reduced cost of treating groundwater to potable standards.

As this empirical study established, there was a secular rise in groundwater levels in the regions which are receiving Narmada waters either in the form of direct delivery through canals in the fields or through discharge of water into the rivers. More importantly, there has been differential rise in water levels during the post-Narmada period, as compared to the pre-Narmada period in many districts, and in one district there has been positive reversal of the water level trend. All the three factors, viz., return flows from surface irrigation, river bed recharge, and reduced groundwater draft for irrigation might have contributed to this change. As a consequence, there has been significant reduction in the TDS in groundwater in all except Surendranagar district. Here also, there has been higher rate of reduction in TDS during the post-Narmada water period as compared to the pre-Narmada period in two districts, namely, Ahmedabad and Kheda, whereas in other districts the rate of reduction in TDS declined. The indirect positive impacts of improved groundwater balance from canal irrigated areas and the dilution of mineral containing groundwater are likely to be very significant in the form of improved irrigation economy based on groundwater, and improved sustainability of rural and urban water supplies based on groundwater.

The improvement in well irrigation economy will result from the reduced cost of pumping groundwater for irrigation owing to reduced pumping depth, reduced incidence of well failures, and increased irrigation potential of wells owing to increase in their yields. Improvement in drinking water economy will result from the reduced cost of pumping water from wells for domestic water supply, the reduced expenditure for the treatment of contaminated groundwater for bringing it to potable standards, and the reduced expenditure for supplying water through tankers in summer months, when wells run dry. We would examine how far these arguments are valid in Chapter 8.

The study reinforces the arguments by water resource scientists, which is fast gaining international recognition, that the water use efficiency in canal irrigation cannot be assessed purely in terms of the total amount of water consumed by the crops in irrigated fields against the total amount of water supplied. They argue that in several instances a significant portion of the applied water, which returns to shallow aquifers or runs off into the drains, is available for reuse downstream (Chakravorty and Umetsu, 2003; Howell, 2001; Seckler 1996). Such engineering notions of water use efficiency lead to the overestimation of the potential for water saving in irrigation (Howell, 2001). Instead, strategies for water use efficiency improvements should look at the total amount of water consumed by the crop, which is beneficial evapo-transpiration, against the total amount of water to minimize the gap between the two (Allen *et al.*, 1998). Well irrigation is sustained in many parts of Punjab and Haryana, Mulla command in Maharashtra, Krishna river delta in Andhra Pradesh and Mahi and Ukai-Kakrapar commands in south Gujarat because of the return flows from surface irrigation (Kumar *et al.*, 2010a).

6

Socioeconomic Impact of Canal Irrigation

Several of the earlier studies reflected on the impact of gravity irrigation from large surface irrigation systems on the farming systems. They broadly covered cropping pattern changes with the introduction of water-intensive crops, increase in input use particularly fertilizers and pesticides, increase in the cost of cultivation, enhancement in yield there by the increase in gross and net income, increase in the cropping intensity, increase in livestock holding, and increase in milk production. Whereas the terrains which are less trucked are the impacts of irrigation on farmers who lift water from canals and the impact of water availability from canals on local irrigation water markets.

In the context of the SSP also, such examination is important because in many villages within the designated command areas, famers were already practicing irrigation either with their own wells or with purchased water from neighboring well owners. Also, the Narmada main canal, and some branch canals and distributaries, currently pass through areas which are either outside the designated command area or where the delivery network is not yet ready for receiving water by gravity. In these areas, farmers are investing their money to lift water from canal. However, these factors make the analysis of the potential impacts less simple and more complex. Adding to this complexity, the SSP covers several agro-climatic regions in Gujarat. Many of these regions are already covered in the areas where the network is completed or where the main and branch canals have already extended. The study region also covers where farmers practice only rain-fed farming, due to absence of groundwater-based sources.

Changes in Water and Agricultural Scenario

Overall Water Resources

In order to see how the water resource scenario in the study region has altered over the recent past, it is important to have quantitative assessment of the allocation of water from the SSP for various sectors of economy. Table 6.1 gives the figures of Gujarat's share of utilizable water resources from the Narmada River basin, in different hydrological years, starting from July 1, 2006, to June 30, 2010. The hydrological year considered here is from July 1 to June 30. It shows that depending on the variation in the total utilizable flow in the basin, Gujarat's share also varied from 13.31 MAF in 2006–2007 to a lowest of 6.14 MAF in 2008–2009. In two of the years, the share was more than the tribunal allocation of 9.75 MAF, while in other two years it was less owing to the reduced yield of the basin. But, the total utilization by Gujarat ranged from 2.06 MAF in 2006–2007 to 5.09 MAF in 2009–2010, with the percentage utilization varying from 18 to 73.10 over the years. The total utilization in two of the four years is slightly higher than the sum of the utilization in three competitive use sectors, viz., irrigation, domestic use, and industry. The excess water might have been used to release into rivers. A closer look at the data clearly points that water utilization for irrigation has been on a steady rise—from 1.00 MAF to 3.41 MAF.

Changes in Agricultural Scenario in the Region

It is undisputable to say that the impact of the SSP on agricultural production and growth has been remarkable. Prior to the introduction of Narmada waters, agricultural growth in Gujarat was highly dependent on groundwater. As per the White Paper on Water in Gujarat prepared

Table 6.1:
Gujarat's Share of the Volumetric Water Allocation from the Narmada River Basin and Utilization over the Years

Period (1 July to 30 June)	Total Utilizable Flow (MAF)	Gujarat's Share of Utilizable Flow (MAF)	Allocation to Various Competitive Use Sectors (MAF)			Total Utilization (MAF)	Percentage Utilization by Gujarat
			Irrigation	Domestic	Industries		
2006–2007	41.42	13.31	1.00	1.03	0.03	2.40	18.00
2007–2008	31.71	10.19	1.91	1.03	0.02	2.96	29.00
2008–2009	19.11	6.14	3.24	0.60	0.05	3.89	63.40
2009–2010	21.66	6.96	3.41	0.85	0.05	5.09	73.10

Source: Data from SSNNL, dated June 30, 2011.

in 2001 (IRMA/UNICEF, 2001), groundwater accounted for nearly 75 percent gross irrigated area in the state, and surface water accounted for the rest (0.77 m. ha). Continuous increase in groundwater irrigation came with a huge social, economic, and environmental cost, with the alluvial aquifers in central and north Gujarat getting mined as a result of average annual water withdrawals far exceeding the annual recharge (Kumar, 2007). The major share of the surface irrigation in Gujarat was concentrated in south Gujarat till the SSP project began delivering irrigation water to central and north Gujarat.

But, the coming of Narmada waters has changed the entire irrigation landscape of Gujarat, with surface irrigation becoming quite prominent in many areas. Satellites imageries processed by Indian Space Research Organization show that as on January 31, 2009, the SSP canals irrigated 0.37 million ha of land in Gujarat. On September 28, 2008, it was 0.25 million hectares. According to another source, the SSP is delivering to farmers, water to irrigate 0.62 million ha of crop land (Alagh, 2010). Though this is only one-third of the designated command area of the SSP, only 25 percent of the distribution system was completed (Kumar et al., 2010b).

Kumar et al. (2010b) showed that the SSP could help Gujarat recover from a major setback in agricultural production after the two consecutive years of drought in 1999 and 2000. The data on gross state domestic product for agriculture (at constant 1999–2000 prices) in Gujarat available for 50 years from 1960–1961 to 2009–2010 show the agricultural growth trend in the state (Department of Economics and Statistics, Government of Gujarat), presented graphically in Figure 6.1. It shows that the agricultural growth in the state follows an exponential trend line ($R^2 = 0.78$). It further shows that there has been a major decline in agricultural outputs in value terms during every severe drought, like in 1985 to 1987 and 1999 to 2000. Further, it shows a gradual recovery from the major dip which occurred in 2000–2001 in the subsequent seven years. This also coincided with the time period, which witnessed gradual expansion in the area irrigated by the SSP.

While the actual area irrigated by canals built under the SSP was nearly 0.4 million ha confined to south Gujarat and central Gujarat, significant results were produced by the water which was carried through

Figure 6.1:
Growth in Agricultural GSDP in Gujarat, Constant Prices (1999–2000)

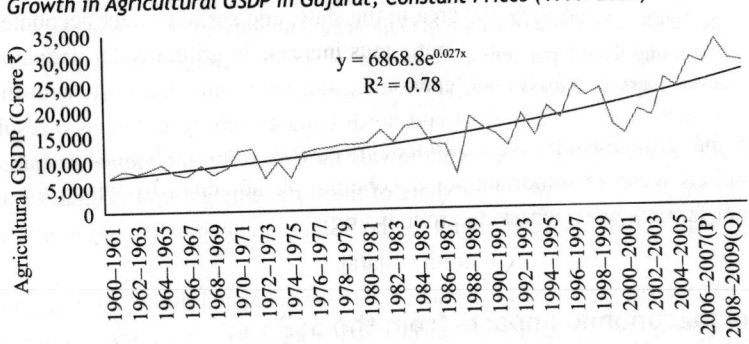

Source: Author.

the Narmada main canal that goes to Kutch and Rajasthan. Farmers, who have agricultural land located on both sides of the main canal and the rivers, pump out water from them to irrigate their fields. It is not uncommon to find farmers who are using rubber pipes as long as 5–6 km to carry water to distant fields. The large investments which farmers made for conveyance of water from the source to their farms show the high value they attach to this new water source.

For the farmers, who have been using groundwater tapped from the deep confined aquifers using expensive tube wells, the availability of surface water became a boon. Besides, being low cost, the quality of water available from canals is very good. Canal water, unlike the groundwater from tube wells is free from salts, contains some micronutrients, and is cooler. It is highly suitable for growing crops such as paddy, banana, and vegetables. In a study looking at the socio-ecological impacts of groundwater degradation in the Sabarmati River basin, Kumar *et al.* found that due to increase in the salinity of tube well water, well irrigators in the alluvial central Gujarat had to abandon cultivation of vegetables (Kumar *et al.*, 2001).

With the introduction of canal water, the farmers in south and central Gujarat have taken up cultivation of irrigated paddy, banana, potato, and other vegetables on a large scale. With the dilution of groundwater in the alluvial aquifers, the well irrigators are also benefited by surface water import. As noted in a National Daily,

Production of the fruit and vegetables registered a robust growth of 12.8 per cent between 2001 and 2008, which is more than double of that in the 1990s. The share of cotton in Gujarat agricultural output has doubled, increasing from 6 per cent to 12 per cent of the gross cropped area within seven years. (*Business Line*, 2009)

In the groundwater overexploited north and central Gujarat, water markets were an important socioeconomic phenomenon by which the poor farmers were accessing groundwater.

Socioeconomic Impacts from the SSP

Changes in Sources of Irrigation and Area under Irrigation

As Table 6.2 shows, the area had some other sources of irrigation, like purchased water, open wells and tube wells/bore wells, prior to the introduction of Narmada canal water. But, the area irrigated by water purchase has reduced after the introduction of Narmada waters through canals in the four districts, with the highest reduction found in Vadodara district which had the highest average area irrigated by farmers with purchased water (0.57 ha) prior to the introduction of Narmada waters. Similarly, well irrigation (either open well or tube well), which the farmers in the command area used to practice in all the four locations in small areas, became nonexistent after the availability of Narmada waters. Overall, as Table 6.2 indicates, area under irrigation increased substantially in all the areas with the introduction of water through Narmada canals. It increased from 1.27 ha to 5.11 ha in Bharuch, 0.1 ha to 0.73 ha in Narmada, 0.42 ha to 0.99 ha in Panchmahals, and 0.78 ha to 1.01 ha in Vadodara district.

The impacts are sharper in the case of canal lift, as found in five locations across Gujarat, viz., Ahmedabad, Bharuch, Narmada, Panchmahals, and Vadodara (Table 6.3). There is a remarkable increase in area irrigated in all locations, from 8.23 ha to 14.17 ha in the case of Ahmedabad; 0.2 ha to 2.89 ha in Narmada; 1.99 ha to 5.68 ha in Vadodara; and 0.0 ha to 3.84 ha in Surendranagar. The increase was, however, very marginal in Mehsana. A detailed analysis of the socio-technical characteristics

Table 6.2:
Changes in Sources of Irrigation and Area Irrigated with the Introduction of Narmada Waters in Canal Command Areas

Name of the Occupation	Name of the Location			
	Bharuch	*Narmada*	*Panchmahals*	*Vadodara*
Before Narmada Water				
Average Irrigated Area (ha per farmer)				
a. Bore well/Tube well	1.04		0.13	0.07
b. Open well	0.06	0.065	0.08	0.14
c. Canal			0.00	0.00
d. Purchased water	0.17	0.034	0.21	0.57
Total from all sources	1.27	0.099	0.42	0.78
After Narmada Water				
Average Irrigated Area (ha per farmer)				
a. Bore well/Tube well				
b. Open well				
c. Canal	5.09	0.72	0.98	1.01
d. Purchased water	0.02	0.011	0.008	
Total from all sources	5.11	0.73	0.99	1.01

Source: Authors' own analysis based on a primary survey.

of canal lift irrigation, for one of the locations, that is, Surendranagar, shows that the investment required for making use of the canal water is often quite large, most of which goes into installing buried pipelines. While the average total capital cost is ₹0.31 million, nearly 88 percent (₹0.28 million) of it is for conveyance system.

In these locations, with the availability of water from canals, the farmers whose land is located on the opposite side of the canals, but falling outside the command, have replaced irrigation by wells and tube wells, with canal water. For instance, in Ahmedabad, the average area irrigated by farmers through tube wells and open wells was 6.8 ha prior to Narmada canal, it got reduced to 0.05 ha per farmer, whereas the average area irrigated through canal lift went up from 0.02 ha to 14.17 ha. Similarly, in Bharuch, the area irrigated by wells and tube wells came down from 6.19 ha to nil, while the canal lift could bring in average 24.91 ha of land. Unlike in other areas, in Surendranagar, the farmers

Table 6.3:
Changes in the Source of Irrigation and Irrigated Area with Irrigation through Canal Lift

Name of the Occupation	Name of the Location					
	Ahmedabad	Mehsana	Narmada	Vadodara	Bharuch	Surendra-nagar
Area irrigated (ha)	**Before Narmada Water**					
Bore well/ Tube well	5.10	2.62		0.83	6.19	
Open well	1.70	0.44		0.79		
Canal	0.02			0.02		
Purchased water	1.41	5.16	0.21	0.35	1.00	
Total from all sources (ha)	8.23	8.22	0.21	1.99	7.19	
Area irrigated (ha)	**After Narmada Water**					
Bore well/ Tube well						
Open well	0.05					
Canal	14.17	8.46	2.89	5.68	24.91	3.84
Purchased water					0.14	
Total from all sources (ha)	14.17	8.46	2.89	5.68	25.05	3.84

Source: Authors' own analysis based on a primary survey.

were largely dependent on rain-fed cultivation prior to Narmada branch canal construction, as groundwater in the area is saline. The most dramatic impact of canal irrigation was, therefore, seen in this area (see the case studies on the impact of canal irrigation in two villages of Limdi taluka of the district).

Equally sharp is the reduction in the area irrigated with purchased water, after the introduction of canals in the area. The average area irrigated with purchased water reduced from 1.4 ha to nil in Ahmedabad, 5.16 ha to nil in Mehsana, 0.21 ha to nil in Narmada district, 0.35 ha to nil in Vadodara district, and 1 ha to 0.14 ha in Bharuch.

Refer to Figure 6.2 for source-wise change in irrigated area (ha).

Figure 6.2:
Source-wise Change in Irrigated Area (ha)

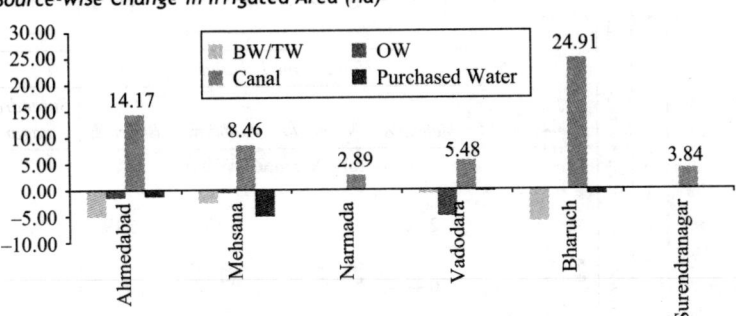

Source: Author.

Changes in Cropping Pattern, Gross Cropped Area, and Cropping Intensity

In the canal command areas, the gross cropped area over the three crops was found to have increased as a result of increase in the availability of cheap water from canals. Farmers were able to grow crops that require irrigation during winter and summer months. The estimated average gross cropped area per farmer increased in the districts from 0.81 ha to 0.848 ha in Panchmahals, 5.09 ha to 6.02 ha in Narmada, and 5.20 ha to 5.64 ha in Vadodara, and in Bharuch, the gross cropped area reduced slightly from 5.21 ha to 5.16 ha. This is not indicative of any reduction in the availability of water. The changes in the area under irrigated crops can only be indicative of the water availability situation. One of the reasons for the absence of expansion in the gross cropped area in Bharuch in spite of an increase in the irrigated area is that the available water is allocated to the long duration varieties of the existing crops such as cotton and castor, which are irrigated. This also constrains the ability of the farmers to expand the cropped area in summer.

Table 6.4 shows the average percentage area under different crops grown by farmers benefited by gravity irrigation from canals before and after the introduction of Narmada waters. With the introduction of Narmada waters, there has been notable change in the cropping pattern in the command area and is characterized by increase in the proportion

Table 6.4:
Percentage Area under Crops in Different Seasons before and after Narmada Waters

Name of Season	Name of Crop	Before Narmada				After Narmada			
		Bharuch	Narmada	Panchmahals	Vadodara	Bharuch	Narmada	Panchmahals	Vadodara
Kharif	Cotton	45.68	54.46	28.09	53.08	53.29	44.81	19.93	42.91
	Chick Pea	40.12	11.14	14.15	10.00	32.25	7.43	7.47	3.19
	Castor	6.14		12.38	3.27	6.59		19.93	7.62
	Green Gram	2.30				2.33			
	Jowar	2.50	0.50		3.65	1.94			
	Sugarcane	1.73				1.74			
	Maize		25.99	14.34			21.04	8.97	
	Paddy		6.19	9.23	14.62		5.90	9.97	17.02
	Groundnut						10.61		
	Cluster bean			0.59				2.49	
	Bajra			4.52				2.82	
Winter	Wheat	1.54	0.50	13.75	1.92	1.55	7.08	18.27	1.77
	Maize		1.24		13.08		3.54	2.49	24.29
	Bajra					0.00		1.66	
	Alfalfa								0.35
Summer	Bajra			2.95	0.38		0.47	7.14	0.89

Source: Authors' own estimates based on a primary survey.

of area under irrigated crops and reduction in area under rain-fed crops. The increased preference for irrigated crops is the incentive of higher income return which they can generate as compared to the rain-fed crops. An example is cotton in Bharuch district, which recorded a significant rise in terms of the proportion of the area under that crop, from 45.7 to 53.3 percent. In Panchmahals and Vadodara districts, however, the proportion of the area under cotton reduced post-Narmada. But, in two of these districts (Panchmahals and Vadodara), the proportion of the area under castor (another irrigated crop) increased substantially. Notably, in these districts, the proportion of rain-fed crops, such as jowar and maize, reduced significantly.

In all locations, the area under winter crops such as wheat and maize increased significantly. For instance, the area under wheat went up from 13.75 percent to 18.3 percent in Panchmahals. Similarly, in Vadodara district, there was substantial increase in area under maize from 13.1 percent to 24.3 percent of the gross cropped area. Area under summer bajra increased in three locations, viz., Narmada, Panchmahals, and Vadodara, an indication of irrigation water availability during that part of the year.

As regards gross cropped area in the areas irrigated by canal lift, the analysis shows the following: In the case of Ahmedabad, the average gross cropped area increased from 5.28 ha to 17.28 ha. In Mehsana, it increased from 2.75 ha to 10.79ha. In Narmada, the highest increase (in percentage terms) was found—with the area increasing from 0.664 ha prior to Narmada to 8.20 ha post-Narmada. In Vadodara, it shot up from 1.36 ha to 7.46 ha, and in Bharuch, it reduced marginally from 5.85 ha to 5.8 ha. As pointed out, the provision of irrigation to the existing long duration crops in gross cropped area in Bharuch in spite of having observed an increase in irrigated area is that the available water is allocated to the long duration varieties of the existing crops such as cotton and castor, which are irrigated. This also limits the ability of the farmers to expand the cropped area of summer. Given the fact that the average area reported to have been irrigated in Bharuch (25.05 ha) is much larger in comparison to the gross cropped area (5.8 ha) post-Narmada, it is quite likely that the irrigated area reported by the farmers also included the area irrigated by water buyers.

Table 6.5 presents shift of the farmers using water from canal through lift in their cropping pattern during the post-Narmada canal period. It shows that the percentage area under crops that are normally rain fed

Table 6.5:
Percentage Area under Crops in Different Seasons before and after Irrigation through Canal Lifting

Name of Crop	Pre-Narmada Canal						Post-Narmada Canal					
	Ahmedabad	Mehsana	Narmada	Vadodara	Bharuch	Surendranagar	Ahmedabad	Mehsana	Narmada	Vadodara	Bharuch	Surendranagar
Kharif												
Cotton	30.1	20.4	48.2	16.9	46.0	63.9	0.3	18.5	4.9	38.5	52.6	
Bt Cotton						1.8						40.2
Chick Pea		9.0		6.7	38.8				48.8	3.9	32.1	
Castor		14.2		6.0	7.4			8.5		16.3	8.1	1.9
Green Gram					2.7						2.8	
Jowar	2.7	5.8	0.0	8.8	3.1	19.7	1.4	5.0		3.0	2.4	17.4
Sugarcane					2.1						2.1	
Maize			33.1	8.8					11.4	3.9		
Paddy	29.5	20.7	6.0	4.6			48.4	25.1	4.9	10.8		
Bajra		1.1				12.3		1.2				6.5
Mustard						0.4						2.5
Groundnut									13.8			
Tomato		16.0						15.0				
Tobacco				44.7						1.3		

Crop										
Winter										
Wheat	37.1	19.3	1.8	3.4	5.7	48.8	22.7	8.1	15.0	5.7
Maize			1.8					8.1	6.0	
Jowar									0.2	
Bajra		0.4							0.6	
Alfalfa	0.4	1.8				0.9	1.5			20.4
Cumin	0.2					0.2	0.2			
Fennel		0.4					1.9			
Mustard					1.4					0.6
Green Gram										1.9
Onion										0.6
Summer										
Bajra							0.4			
Groundnut									0.5	

Source: Authors' own analysis based on data collected through a primary survey in the five locations.

reduced in all the locations. For instance, the area under jowar, as a percentage of the gross cropped area, reduced in four locations. Similarly, the percentage area under kharif maize also reduced in the two locations where it was grown. On the other hand, there was a notable increase in the area under kharif paddy, which is grown with supplementary irrigation, in three locations, viz., Ahmedabad, Mehsana, and Vadodara, though similar proportion of the area under this crop reduced in Narmada district. As regards cotton, there was a marked reduction in the proportion of area under this crop in Ahmedabad and Narmada, whereas in Vadodara and Bharuch, it increased substantially. In Surendranagar, the area under rain-fed cotton reduced drastically, but there was a sharp increase in area under BT cotton—from 1.8 percent to 40.2 percent. As regards castor, the percentage area under this crop increased in Vadodara, while it reduced in Mehsana. In Narmada district, there was a substantial increase in the area under chick pea from 9 percent to 48.8 percent.

What is most striking is the tendency of canal irrigators to allocate greater proportion of their land to irrigated winter crops. A substantial increase in irrigated wheat was noted in all the four locations. As regards winter maize, the crop was newly introduced by the farmers interviewed in the study in Vadodara district command. Besides, the area under this crop as a percentage of the gross cropped area has become quite substantial in both Narmada and Vadodara districts. Cumin was newly introduced in Surendranagar, with the crop occupying more than 20 percent of the gross cropped area.

Changes in Yield and Returns of Crops Grown in Different Seasons

One of the widely visible first-order impacts of irrigation is on crop yields. The impacts can be due to the direct effect of irrigation water on crop yields (through the biophysical changes) or the impact of improved irrigation on crop technology, which is the impact on total factor productivity (Evenson *et al.*, 1999). The yields of crops grown in the Narmada canal commands in four locations pre- and post-Narmada water introduction are presented in Table 6.6. The graphical representation of the change in the yield of various crops across seasons after the introduction of gravity irrigation from Narmada is provided in Figure 6.3.

Table 6.6:
Change in Crop Yield (kg per ha) under Canal Irrigation

Name of Season	Name of the Crops	Crop Yield in			
		Bharuch	*Narmada*	*Panchmahals*	*Vadodara*
Before Narmada					
Monsoon	Cotton	997.9	1,078.7	1,206.9	1,295.1
	Chick Pea	664.3	553	603.5	832.2
	Castor	919.8		2,482.0	934.4
	Green Gram	876			
	Jowar	1,314	1,226.4		1,296.5
	Sugarcane	17,520			
	Maize		1,713.1	2,055.2	2,190.0
	Paddy		1,576.8	1,423.5	2,236.4
	Groundnut				
	Bajra			2,299.5	
	Cluster Bean			876.0	
Winter	Wheat	2,190.0	2,628.0	2,244.8	1,255.6
	Maize		1,861.5		2,136.1
	Jowar				
Summer	Bajra			1,606.0	1,752.0
After Narmada					
Monsoon	Cotton	2,156.3	1,869.4	1,927.2	2,099.8
	Chick Pea	1,101.9	1,188.9	832.2	1,423.5
	Castor	2,628.0		2,555	2,277.6
	Green Gram	1,029.3			
	Jowar	1,752.0			
	Sugarcane	26,280.0			
	Maize		2,886.8	2,514.1	
	Paddy		2,815.7	2,102.4	3,698.7
	Groundnut		2,080.5		
	Bajra			3,066.0	
	Cluster Bean			788.4	
Winter	Wheat	3,504.0	3,230.3	2,839.7	2,978.4
	Maize		2,645.5	2,487.8	3,975.7
Summer	Bajra		4,380.0	3,153.6	4,380.0

(Table 6.6 Contd)

(Table 6.6 Contd)

Name of Season	Name of the Crops	Crop Yield in			
		Bharuch	*Narmada*	*Panchmahals*	*Vadodara*
Change in the Crop Yield					
Monsoon	Cotton	1,158.4	790.7	720.3	804.7
	Chick Pea	437.6	635.9	228.7	591.3
	Castor	1,708.2		73.0	1,343.2
	Green Gram	153.3			
	Jowar	438.0	−1226.4		−1,296.5
	Sugarcane	8,760.0			
	Maize		1,173.7	458.9	−2,190.0
	Paddy		1,238.9	678.9	1,462.3
	Groundnut		2,080.5		
	Bajra			766.5	
	Cluster Bean			−87.6	
Winter	Wheat	1,314.0	602.3	594.9	1,722.8
	Maize		784	2,487.8	1,839.6
	Jowar			1,752.0	
Summer	Bajra		4,380.0	1,547.6	2,628.0

Source: Authors' own estimates based on primary data from canal command area.

Figure 6.3:
Impact of Narmada on the Yield of Crops (kg per ha) Irrigated by Canal (Kharif)

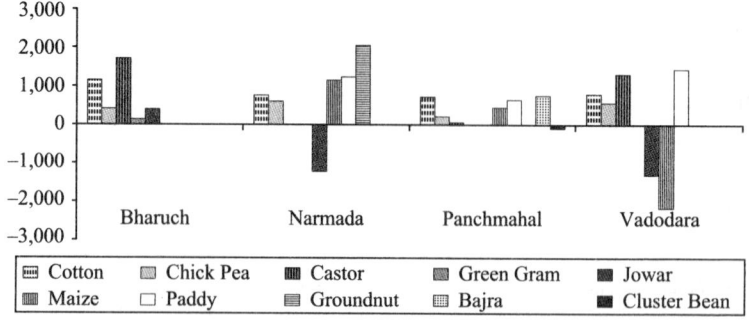

Source: Author.

Table 6.6 shows that there is a remarkable increase in the yield of crops. This is quite notable. Like variation in yields, the extent of rise in yield resulting from canal water introduction varies from location to location for the same crop. In the case of cotton, the yield increase ranged from 59.7 percent to 116 percent. The highest increase in cotton yield was found in Bharuch, which is known for the black cotton soils. As seen in Table 6.5, the command area in the district also experienced a shift in cropping pattern toward cotton. In the case of castor, the yield increase ranged from 3 percent to 185 percent. Here again, the highest rise in the yield was found in the case of Bharuch. For paddy, the rise was in the range 47.7 percent–78.6 percent. The yield increase was found even for crops which are not irrigated under normal circumstances.[1] For wheat, the extent of increase ranged from 22.9 percent to 137.2 percent.

Table 6.7 presents the results vis-à-vis the crop yield pre- and post-Narmada and change in the yield of crops post-Narmada water introduction for canal lift irrigators in six locations. Here again, the change in the yield has been positive everywhere, indicating increase in yield throughout the locations and for all the crops, which are grown by the farmers. The yield improvement is substantial for many crops, including paddy, wheat, and cotton. In the case of wheat and paddy in Narmada district, the increase was phenomenal, touching 2.4 ton per ha for wheat and 3.06 ton per ha for paddy. In Surendranagar district, the wheat yield increased from 1,314 kg per ha to 2,010 kg per ha. But, here, the farmers used to grow rain-fed wheat before Narmada, and this used to fetch a premium price of ₹19 per kg. But, the wheat which is grown these days is irrigated wheat, which fetches a much lower price. In the case of cotton in Mehsana, it was 2.48 ton per ha. In Surendranagar, cotton recorded an impressive yield from 876 kg per ha to 1,248 kg per ha, an increase of 368 kg per ha of land. The graphical representation of the change in the yield of various crops across seasons after the introduction of canal water by farmers through lift is provided in Figure 6.4.

The yield improvement could be attributed to the following five factors. First, availability of canal water, which is cheap, might be the

---

[1] Normal circumstance here refers to situations where the rainfall in the area is normal, wherein the effective rainfall would be sufficient to take care of the crop water requirements.

Table 6.7:
Change in Yield (kg per ha) of Crops with the Introduction of Water from Narmada Canal through Lifting

Name of the Season	Name of the Crops	Name of the Location					
		Ahmedabad	Mehsana	Narmada	Vadodara	Bharuch	Surendranagar
Before Narmada							
Monsoon	Cotton	1,109.6	2,028.49	957.76	1,648.94	968.75	734.3
	BT Cotton						876.0
	Chick Pea			635.1	621.96	814.16	
	Castor		1,873.67		1,635.2	963.6	
	Green Gram					934.4	
	Jowar	1,533.0			1,752.0	1,095.0	
	Sugarcane					35,040.0	
	Maize			1,603.08	1,106.95		
	Paddy	2,447.59	3,293.11	876.0	3,358.0		
	Groundnut						
	Bajra		1,270.2				1,059.2
	Mustard						438.0
	Tomato		34,622.9				
	Tobacco				1,676.91		
Winter	Wheat	2,365.2	2,671.8	1,314.0	2,409.0		1,314.0
	Maize			1,693.6	2,409.0		
	Jowar						
	Green Gram						1,752.0

Season	Crop						
	Cumin	350.4	613.2				
	Fennel		1,007.4				
Summer	Jowar		876.0				
	Groundnut				1,752.0		
After Narmada							
Monsoon	Cotton	1,752.0	4,509.78	1,560.38	1,738.52	2,259.16	
	BT Cotton			876.0	1,427.88	1,078.15	1,248.8
	Chick Pea		2,197.3		2,092.67	2,628.0	1,314.0
	Castor					1,423.5	
	Green Gram						2,487.2
	Jowar	2,628.0	2,092.67			1,533.0	
	Sugarcane					43,800.0	
	Maize			3,504.0	2,084.88		
	Paddy	4,077.93	3,641.66	3,942.0	3,744.9		
	Groundnut	1,752.0		2,365.2			
	Bajra		1,752.0				1,757.48
	Mustard						700.8
	Tomato		40,671.4		39,402.0		
	Tobacco						

(Table 6.7 Contd)

(Table 6.7 Contd)

Name of the Season	Name of the Crops	Name of the Location					
		Ahmedabad	Mehsana	Narmada	Vadodara	Bharuch	Surendranagar
Winter	Wheat	4,186.98	3,033.15	3,723.0	3,040.24		2,010.0
	Maize			2,890.8	2,468.73		
	Jowar				3,066.0		
	Green Gram						1,752.0
	Cumin	613.2					587.4
	Fennel		1,182.6				
	Onion						1,752.0
	Mustard						350.4
Summer	Jowar						
	Groundnut				2,190.0		
Change in Crop yield (kg per ha)							
Monsoon	Cotton	642.4	2,481.29	602.62	89.58	1,290.41	
	BT Cotton						
	Chick Pea			240.9		263.99	368.8
	Castor		323.63		805.92	1,664.4	1,314.0
	Green Gram				457.47	489.1	
	Jowar	1,095.0	2,092.67		-1,752.0	438.0	
	Sugarcane					8,760.0	
	Maize			1,900.92	977.93		2,487.2
	Paddy	1,630.34	348.55	3,066.0	386.9		

Season	Crop					
	Groundnut	698.32	2,365.2			
	Bajra	262.8			481.8	
	Mustard				6,048.5	
	Tomato			2,265.09		
	Tobacco	696.0				
Winter	Wheat		2,409.0	631.24	361.35	1,821.78
	Maize		1,197.2	59.73		
	Jowar			3,066.0		
	Green Gram	0				
	Cumin	587.4			−613.2	262.8
	Fennel				175.2	
	Onion	1,752.0				
	Mustard	350.4				
Summer	Jowar				−876	
	Groundnut			438		

Source: Authors' own analysis based on data collected through a primary survey in the five locations.

Figure 6.4:
Impact of Narmada on Yield of Crops (kg per ha) Irrigated through Canal Lift (Kharif)

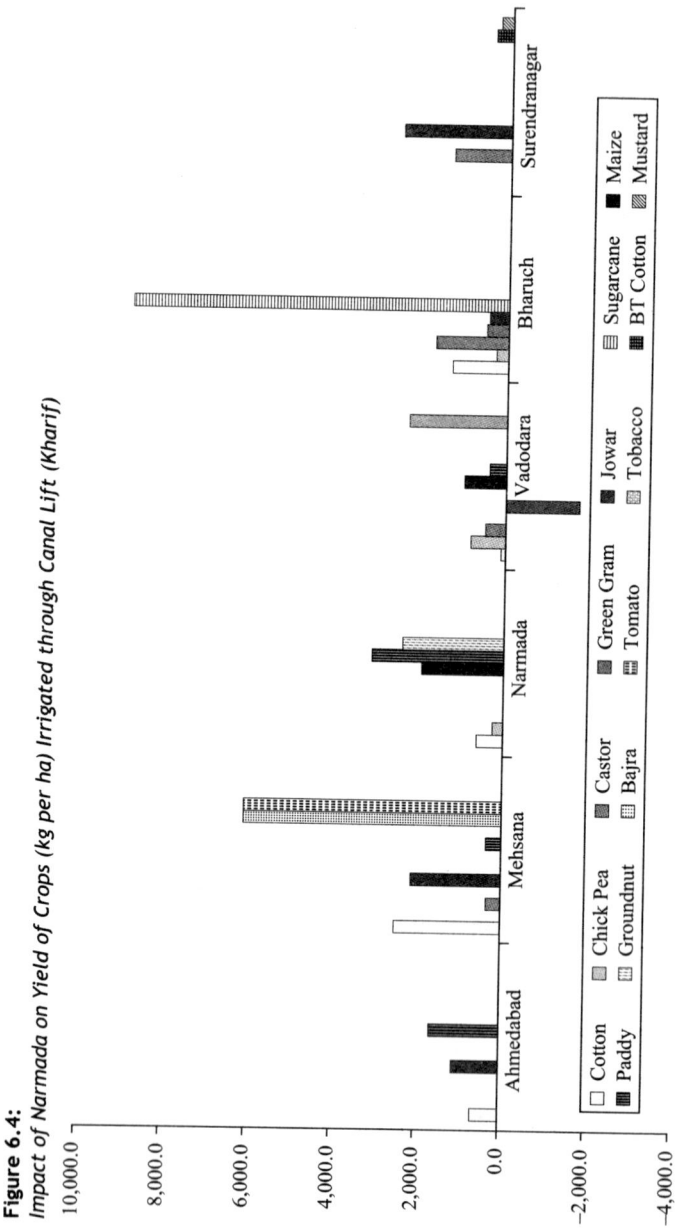

Source: Author.

Figure 6.5:
Number of Rain-fed and Irrigated Crops before and after Narmada Waters (Canal Command Area)

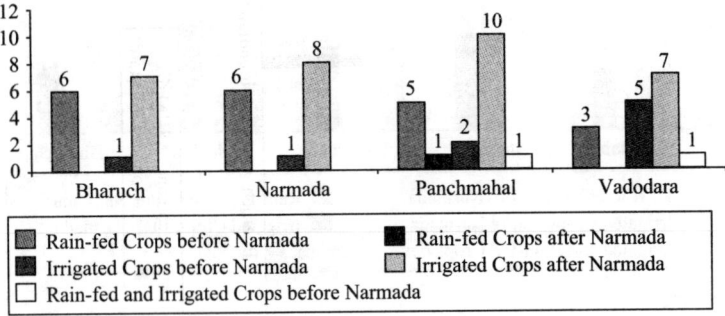

Source: Author.

reason which motivated farmers to provide optimum irrigation to the crop to maximize the yield, which would not have been possible with purchased water or water lifted from wells using electricity or diesel as they involve substantial costs. This is a direct impact of irrigation. Second, the crops which are normally raised under rain-fed conditions such as cluster bean, paddy, jowar, chick pea, etc., are now provided with supplementary irrigation, thereby resulting in yield enhancement, with simultaneous replacement of short duration varieties of some crops with their log duration counterparts.[2]

The change in the number of irrigated crops due to the introduction of canal water in canal command areas and area irrigated through canal lift are given in Figures 6.5 and 6.6. As Figure 6.5 shows, there has been substantial increase in the number of irrigated crops after Narmada canal water introduction in all the four locations (1–7 in Bharuch, 1–8 in Narmada, 2–10 in Panchmahals, and 5–7 in Vadodara). Unlike, the three other locations, in Vadodara, majority of the crops were irrigated even prior to the introduction of Narmada canal water. Further, in none of the locations, farmers grow crops under rain-fed conditions post-Narmada

---

[2] Cotton and castor are good examples. In Surendranagar, it was found that after Narmada water was made available, the farmers in the area, who used to grow rain-fed cotton, started growing only long-duration, hybrid varieties of cotton, which required irrigation, but gave higher yield.

Figure 6.6:
Number of Rain-fed and Irrigated Crops before and after Narmada Canal Lift

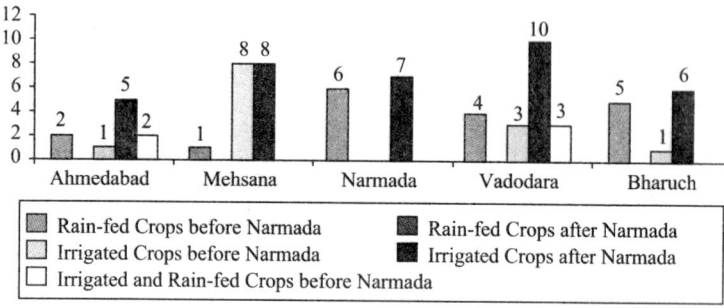

Rain-fed Crops before Narmada   Rain-fed Crops after Narmada
Irrigated Crops before Narmada   Irrigated Crops after Narmada
Irrigated and Rain-fed Crops before Narmada

Source: Author.

waters. Similar trend vis-à-vis increase in the number of irrigated crops was seen with farmers irrigating crops through canal lift. In this case, even prior to the introduction of canal water, farmers used to grow a few crops under irrigated conditions in four out of the five locations. It was only in Narmada that farmers were growing crops under rain-fed conditions prior to Narmada waters.

The third reason for the yield improvement is that canal water is generally of good quality and contains certain micronutrients, unlike groundwater, which contains salts that can often harm the crops, (Kumar and Amarasinghe, 2009, based on a study conducted in the canal-irrigated areas of Punjab). Fourth: The very fact that the availability of irrigation water had enabled many farmers to take irrigated crops—which was manifested in the expansion in average area under irrigated crops per farmer—such as cotton, castor, wheat, maize, and sugarcane that give high returns per unit of land, in kharif, winter, and sometimes summer, resulting in greater farm surplus from the same amount of land, had increased their ability to invest in agronomical inputs, including high-quality seeds of high-yielding varieties, labor, and fertilizers and pesticides even for rain-fed crops. Fifth: Finally, the increased availability of irrigation water, which comes with greater dependability, also increases the confidence of farmers to take greater risk and make more investments in crop inputs.

This in fact is the positive externality of irrigated agriculture. It is reflected in the cost of cultivation of the crops (Tables 6.8 and 6.9 for

Table 6.8:
The Cost of Cultivation of Canal Irrigators (₹ per ha)

Name of the Season	Name of the Crops	Bharuch	Narmada	Panchmahals	Vadodara
Before Narmada					
Kharif	1. Cotton	15,596.61	15,368.2	15,963.6	13,331.8
	2. Chick Pea	9,072.08	7,072.3	8,398.4	8,442.5
	3. Castor	12,154.50		12,483.0	11,504.8
	4. Green Gram	8,541.00		0.0	
	5. Jowar	8,760.00	10,095.9	0.0	6,920.4
	6. Sugarcane	16,425.00		0.0	
	7. Maize		11,036.0	13,305.1	10,161.6
	8. Paddy		13,928.4	18,231.8	14,461.7
	10. Bajra			11,311.4	
	11. Cluster bean			3,504.0	
Winter	1. Wheat	10,205.40	14,454.0	12,748.5	9,767.4
	2. Maize		10,304.0		13,941.9
Summer	1. Bajra			10,081.3	7,971.6
After Narmada					
	1. Cotton	32,263.8	33,879.3	30,222.0	25,162.6
	2. Chick pea	19,876.0	22,056.4	15,695.0	13,260.5
	3. Castor	31,974.0		29,141.6	31,124.3
	4. Green gram	24,745.1			
	5. Jowar	11,716.5			
	6. Sugarcane	36,354.0			
	7. Maize		22,094.1	5,518.8	
	8. Paddy		19,534.8	19,140.6	25,670.5
	9. Groundnut		37,543.9		
	10. Bajra			19,607.8	
	11. Cluster bean			7,665.0	
Winter	1. Wheat	19,710.0	22,715.8	19,987.4	14,862.8
	2. Maize		20,169.9	16,906.8	24,378.1
	3. Bajra			91,98.0	15,877.5
Summer	1. Bajra		14,782.5	20,121.7	11,607.0
Change in the Cost of Cultivation					
	1. Cotton	16,667.2	18,511.1	14,258.4	11,830.8
	2. Chick pea	10,803.9	14,984.1	7,296.6	4,818

(Table 6.8 Contd)

(Table 6.8 Contd)

Name of the Season	Name of the Crops	Cost of Cultivation of the Crop in			
		Bharuch	*Narmada*	*Panchmahals*	*Vadodara*
	3. Castor	19,819.5		16,658.6	19,619.5
	4. Green gram	16,204.1			
	5. Jowar	2,956.5	–10,096		–6,920.4
	6. Sugarcane	19,929			
	7. Maize		11,058.1	–7,786.3	–10,162
	8. Paddy		5,606.4	908.8	11,208.8
	10. Bajra		37,543.9	–11,311	
	11. Cluster bean			16,103.8	
Winter	1. Wheat	–10,205.4	–14,454	–5,083.5	–9,767.4
	2. Maize	19,710	12,411.8	19,987.4	920.9
Summer	1. Bajra	0	20,169.9	6,825.5	16,406.5

Source: Authors' own estimates based on data collected through primary survey in four locations.

irrigators in canal commands and the farmers who are lifting water from the canals, respectively). For almost all the crops, the cost of cultivation has increased after the introduction of Narmada waters. The percentage increase varies from crop to crop and location to location, from as high as 120.5 for cotton in Narmada district to 170 percent for castor in Vadodara district to 211 percent for Chick pea in Narmada district.

Changes in Income from Crop Production

We have already seen that the crop yields increased considerably across the crop types grown in different seasons after the introduction of canal water. Higher yield means higher gross income, even if the prices have remained the same. Ideally, the prices of agricultural commodities (cereals, pulses, fiber, oil seeds, and vegetables) also might have increased over time due to inflation. But, we have also seen that the average cost of inputs increased substantially under the canal irrigation scenario for all the crops. The increase in input prices could be probably because of the combined effect of: (1) increase in quantum of inputs such as fertilizers and pesticides and labor inputs, and improved quality

Table 6.9:
Change in the Cost of Cultivation (₹ per ha) of Crops Irrigated through Canal Lifting

Name of the Season	Name of the Crops	Cost of Cultivation in					
		Ahmedabad	Mehsana	Narmada	Vadodara	Bharuch	Surendranagar
Before Narmada							
Monsoon	Cotton	6,943.1	24,857.9	17,013.4	16,700.7	16,167.4	8,707.13
	BT Cotton						16,245.42
	Chick Pea			7,117.5	8,718.8	11,549	
	Castor		20,073.8		17,520.0	12,154.5	
	Green Gram					12,775.0	
	Jowar	3,832.5			10,804.0	13,140.0	
	Sugarcane					24,528.0	
	Maize			11,804.1	10,119.8		
	Paddy	17,010.3	22,156.1	12,548.7	15,622.0		
	Groundnut						
	Bajra		9,395.1				4,633.34
	Mustard						6,153.9
	Tomato		43,115.9				
	Tobacco				15,497.9		
Winter	Wheat	15,013.7	15,543.8	10,950.0	17,301.0		5,913
	Maize			12,395.40	12,088.8		
	Bajra						

(Table 6.9 Contd)

(Table 6.9 Contd)

Name of the Season	Name of the Crops	Cost of Cultivation in					
		Ahmedabad	Mehsana	Narmada	Vadodara	Bharuch	Surendranagar
	Green Gram						5,343.6
	Cumin	24,309.0	21,462.0				
	Fennel		14,716.8				
	Onion						
	Mustard						
Summer	Bajra						
	Jowar		7,029.90				
	Groundnut				14,673.00		
After Narmada							
Monsoon	Cotton	24,703.2	40,092.1	34,232.4	29,408.9	33,315.7	
	BT Cotton						30,015
	Chick Pea			19,272.0	17,671.1	20,420.9	
	Castor		29,459.2		24,832.2	31,974.0	38,303
	Green Gram					18,779.3	
	Jowar	12,351.6	17,797.4			14,782.5	11,230.7
	Sugarcane					36,354.0	
	Maize			27,002.7	20,200.6		
	Paddy	26,046.0	30,024.9	21,286.8	27,514.6		
	Groundnut			39,428.8			
	Bajra		13,490.4				10,133.4

Season	Crop						
	Mustard						11,129.5
	Tomato		52,950.0				
	Tobacco				41,172.0		13,550
Winter	Wheat	21,209.8	22,227.4	21,724.8	21,565.1		
	Maize			27,156.0	18,232.7		
	Bajra				19,929.0		11,300.0
	Green Gram						31,370.4
	Cumin	55,713.6					
	Fennel		22,425.6				12,614.0
	Onion						12,912.0
	Mustard						
Summer	Bajra		13,140.0				
	Jowar				21,462.0		
	Groundnut						
Change in the Cost of Cultivation							
Monsoon	Cotton	17,760.09	15,234.2	17,219	12,708.2	17,148.31	
	BT Cotton						13,769.5
	Chick Pea			12,154.5	8,952.27	8,871.88	
	Castor		9,385.4		24,832.2	19,819.5	38,303.1
	Green Gram					6,004.25	
	Jowar	8,519.1	17,797.4		-10,804	1,642.5	11,230.8
	Sugarcane					11,826	

(Table 6.9 Contd)

(Table 6.9 Contd)

Name of the Season	Name of the Crops	Cost of Cultivation in					
		Ahmedabad	Mehsana	Narmada	Vadodara	Bharuch	Surendranagar
	Maize			15,198.6	10,080.8		
	Paddy	9,035.7	7,868.8	8,738.1	11,892.6		
	Groundnut			39,428.8			
	Bajra		4,095.3				5,500.1
	Mustard						4,975.6
	Tomato		9,834.1				
	Tobacco				25,674.1		
Winter	Wheat	6,196.1	6,683.6	10,774.8	4,264.1		7,637.0
	Maize			14,760.6	6,143.9		
	Bajra				19,929		
	Jowar						
	Green Gram						
	Cumin	31,404.6	−21,462				5,956.8
	Fennel		7,708.8				31,370.4
	Onion						12,614.0
	Mustard						12,912.2
Summer	Bajra						
	Jowar		6,110.1				
	Groundnut				6,789		

Source: Authors' own estimates based on data collected through a primary survey in five locations.

of the seeds used; and (2) the simple inflationary effects on the price of these inputs with time and the removal of subsidies for inputs such as fertilizers. The extent of increase in gross income (as a result of price rise and yield improvement) and cost of inputs would finally determine to what extent the net income for a given crop would change, and whether upwards or downwards.

The average farm gate prices of crops grown by farmers currently in canal irrigated areas in different locations are presented in Table 6.10.

Table 6.11 shows the change in the average net income per ha of cultivated area for 15 crops grown by the farmers in the command area in four different locations. As can be seen from the table, except jowar in Narmada district, paddy in Narmada and Panchmahals district, and kharif maize in Vadodara district, the net income is estimated to have increased for all the crops and in all four locations, wherever such comparisons are possible. It should be mentioned here that in many situations, the size of the sample of farmers who were growing the crop before and after the introduction of canal water, was too small, with the result that

Table 6.10:
Average Farm Gate Price of Different Crops (₹ per kg) in the Canal Command Areas

| Name of the Crop | Farm Gate Prices of Different Crops in Different Locations | | | |
	Bharuch	Narmada	Panchmahals	Vadodara
Cotton	53.2	49.5	46.5	48.7
Tur	33.6	32.7	33.3	34.0
Castor	50.0	–	42.5	39.8
Green Gram	40.6	–	–	–
Sugarcane	2.5	–	–	–
Jowar	10.0	–	10.0	7.5
Wheat	11.0	10.2	10.2	10.8
Maize (Kharif)	–	10.1	10.1	–
Maize (Winter)	–	10.0	10.5	10.6
Paddy	–	10.6	10.0	10.2
Groundnut	–	29.6	–	–
Bajra (Summer)	–	10.5	10.8	10.0
Bajra (Kharif)	–	–	10.5	–
Cluster bean	–	–	21.5	–

Table 6.11:
Changes in Income from Various Crops in Canal Command Area after the Introduction of Narmada Waters

Name of the Season	Name of the Crops	Changes in Income from Crop Production (₹ per ha) for Various Crops in			
		Bharuch	*Narmada*	*Panchmahals*	*Vadodara*
Monsoon	Cotton	67,300.7	51,958.2	49,568.0	69,399.7
	Chick Pea	12,307.8	7,285.0	10,174.7	27,035.6
	Castor	94,279.5		34,988.9	47,686.5
	Green Gram	14,109.1			
	Jowar	5,070.0	−7,540.9		−3,661.7
	Sugarcane	19,491.0			
	Maize		6,791.7	22,816.4	−8,453.4
	Paddy		−14,498.9	−19,009.2	17,978.5
	Groundnut		24,049.9		
	Bajra			15,373.8	
	Cluster Bean			−350.4	
Winter	Wheat	14,848.2	4,741.4	4,505.0	15,052.6
	Maize		6,327.3	13,928.4	19,238.3
	Jowar			19,272.0	23,104.5
Summer	Bajra		32,521.5	23,400.9	21,155.4

Source: Authors' own estimation based on primary data on crop inputs and outputs.

both pre- and post-data are simultaneously not available for analyzing the income changes. The income (net) increase was found to be highest for cotton, across the locations, ranging from a lowest of ₹49,568 per ha in Panchmahals to a highest of ₹69,399 per ha in Vadodara district. An exceptional increase in net income was, however, found in the case of castor in Bharuch district (₹94,279 per ha), which was not visible in the two other locations where the crop was grown. For crops such as green gram and millet, the incremental income was not high. Intriguingly, a remarkable rise in income of up to ₹32,521 per ha was found in the case of summer bajra in Narmada district.

What is more striking is the fact that even for wheat, which was an irrigated crop prior to Narmada water introduction, farmers secured a higher net return, with a lowest increase of ₹4,505 per ha in the case of Panchmahals district to a highest increase of ₹15,052 in the case of Vadodara district. The increase in net income was ₹14,848 per ha for Bharuch. Such a big gain has been made mainly through yield improvement.

Now the income change can occur as a result of change in farm gate price per kg of crop output, change in input costs, in addition to the change in crop outputs per ha. But, what appears as income rise may not be the actual rise in income in real terms. Inflation will impact the net income—calculated at current prices—from crop production due to its effect on the price of crop outputs and inputs. The effect of inflation on the net income is not factored in here while estimating the income change due to the reason that the pre-Narmada farming situation in different locations corresponds to different years and within the same location was also found to be changing. Nevertheless, the effect of inflation on the change in the net income from crops is not expected to be high, as the time lag between pre- and post-Narmada situations ranges from a minimum of two years in most locations to a maximum of seven years in a few locations.

The average farm gate prices of crops grown by farmers lifting canal water for irrigation presently in different locations are presented in Table 6.12.

Table 6.13 shows the estimates of the change in the average net income per ha of the cultivated area of farmers irrigating through canal lifting. The estimated increase in income was found to be very high for cash crops, viz., cumin, fennel, castor, and cotton. The income increase was ₹70,977 per ha for cotton in Bharuch, followed by ₹34,905 per ha for the same crop in Ahmedabad. Again, the increase in the net income was ₹93,622 per ha for castor in Bharuch. For fennel grown in Mehsana, it was ₹55,363 per ha. For cumin in Surendranagar, which was introduced after the arrival of canal water, the farmers started earning ₹49,350 per ha on average. Income increase was also visible for wheat, which was irrigated even prior to Narmada, in four out of the five locations where it was grown.

Here, unlike what was found in the case of canal irrigators, the income has not risen for all the crops and for all the locations. In 5 out of the 45 cases, there has been a decline in the net income post-Narmada waters. The first one of these was for castor in Vadodara (₹17,182 per ha), the second one for paddy in Narmada (₹2,168 per ha), third one for kharif bajra in Mehsana (₹62,130 per ha), the fourth one for winter maize

Table 6.12:
Average Farm Gate Prices of Different Crops (₹ per kg) in Areas Irrigated by Canal Lift

Name of the Crop	Farm Gate Prices of Different Crops in Different Locations				
	Ahmedabad	Mehsana	Narmada	Vadodara	Bharuch
Cotton	35.0	43.6	40.6	45.7	52.6
Jowar	10.0	9.86	–	–	20.0
Paddy	9.7	10.5	11.3	10.2	–
Bajra	–	8.50	–	10.0	–
Maize (K)	–	–	10.40	10.7	–
Tur	–	–	35.00	36.5	33.5
Tomato	–	13.7	–	–	–
Groundnut	–	–	30.00	–	–
Tobacco	–	–	–	17.5	–
Castor	–	39.0	–	35.6	50.0
Green Gram	–	–	–	–	25.0
Wheat	11.0	11.1	10.00	11.2	–
Cumin	150.0		–	–	–
Fennel	–	100.0	–	–	–
Mustard	–	25.0	–	–	–
Maize (Winter)	–	–	9.80	10.6	–
Jowar (Summer)	–	10.0			–
Groundnut	–	–	–	40.0	–
Sugarcane	–	–	–	–	1.50

Source: Author.

in Vadodara (₹11,469 per ha), and the fifth one for cumin in Mehsana (₹55,188 per ha). While this is not explained by the change in the yield of these crops, which has been positive for all crops, such trends could only be explained by the disproportionately higher increase in input costs, which could not be offset by the yield improvements and the price increases. Further, in such instances, the area under the crop has been very negligible (for instance, bajra and cumin in Mehsana), pushing the cost per ha figures upwards, even with very small reduction in the aggregate net income from the crop.

It is to be kept in mind here that unlike in the case of gravity irrigation from canals, irrigation through canal lifting would involve major

Table 6.13:
Change in the Average Net Income of Canal Lift Irrigators after the Introduction of Narmada Waters

Name of the Season	Name of Crop	Change in Average Net Income (₹ per ha) of Farmers in					
		Ahmedabad	Mehsana	Narmada	Vadodara	Bharuch	Surendranagar
Kharif	Cotton	34,905.4	83,633.0	25,675.8	31,163.5	70,977.8	21,703.4
	BT Cotton			8,212.5	31,256.1	11,544.7	7,686.9
	Chick Pea		25,612.0			93,622.5	
	Castor				-17,182.3		13,101.7
	Green Gram	9,657.9				11,680.0	
	Jowar					22,776.0	
	Sugarcane			18,417.9	5,649.4	1,314.0	
	Maize	6,252.8	8,088.4				
	Paddy			-2,168.1	26,600.3		
	Groundnut						2,276.0
	Bajra		-62,130.3				19,114.3
	Mustard						
	Tomato		391,145.0		20,493.2		
	Tobacco						
Winter	Wheat	17,282.94	12,467.01	13,315.2	792.3		10,637.3
	Maize				-11,469.6		
	Bajra			6,278.0			
	Green Gram						7,183.2

(Table 6.13 Contd)

(Table 6.13 Contd)

Name of the Season	Name of Crop	Change in Average Net Income (₹ per ha) of Farmers in				
		Ahmedabad	Mehsana	Narmada	Vadodara	Bharuch	Surendranagar
	Cumin	25,535.4	−55,188.0				49,350.3
	Fennel		55,363.2				
	Onion						4,905.6
	Mustard						3,906.9
Summer	Bajra						
	Jowar		19,512.9				
	Groundnut				45,771.0		

Source: Author.

costs (both capital and variable), depending on the location of the farm vis-à-vis the canal, and whether the canal is in cutting or banking. If the canals are in deep cutting, then lifting would be large, raising the cost of pump operation.

The estimates of the net income being earned by farmers who are using gravity irrigation from canals and canal lift during kharif season are presented in graphical form in Figures 6.7 and 6.8, respectively. Similar estimates for winter and summer crops are presented in Figures 6.9 and 6.10, respectively.

Farming Enterprise in the Non-command Area

As we have discussed in the methodology, one of the approaches to quantitatively assess the impact of irrigation on farming enterprise within the command area is to compare the inputs and outputs of the farmers receiving canal water to those who are outside the command area and not receiving canal water. But, while doing such comparisons, it is important to make sure that the farmers who are compared have more or less identical characteristics in terms of farm holding size and livestock holding, productivity of the farm land, access to irrigation facilities, and domestic labor available at the disposal of the farmers, to engage in farming operations, before canal water was introduced in the area, and the access of the non-command farmers to irrigation facilities has not changed during the period of investigation. Otherwise, such comparisons will not make any sense. Generally, it is found that the land receiving canal water is at lower elevation, as compared to the land outside the command, and also the command area land becomes wetland after some years of receiving water from the canal. Because of this, the kind of crops which command area farmers can take up will be very different from those which farmers in the non-command area take up.

We have painstakingly picked up farmers from three locations in the command area (with 30 farmers from each location), whose land is located outside the canal command, but have more or less similar socio-economic characteristics. The locations are Narmada, Panchmahals, and Vadodara. In addition, nearly 30 farmers who are not lifting water from Narmada canal in Mehsana district were also surveyed for comparing

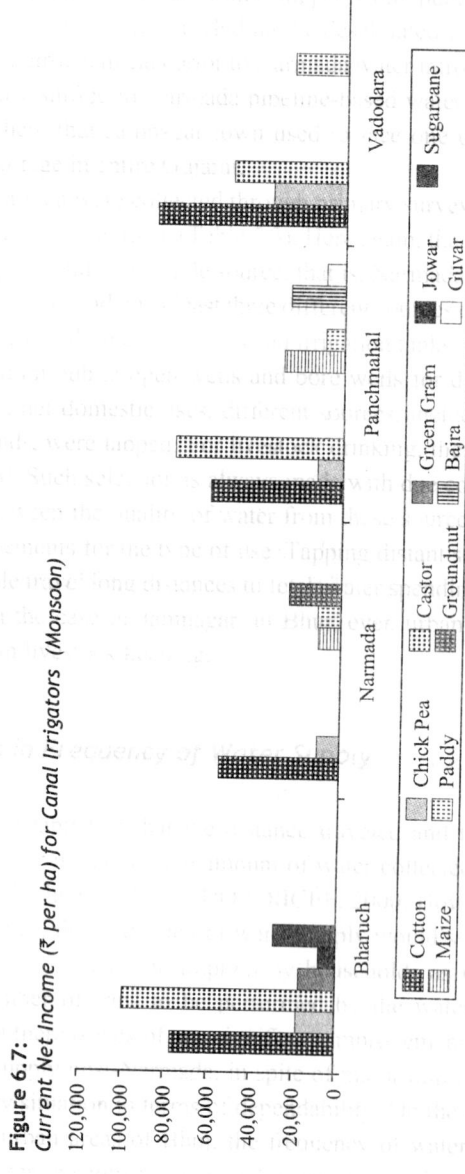

Figure 6.7:
Current Net Income (₹ per ha) for Canal Irrigators (Monsoon)

Source: Author.

Figure 6.8:
Current Net Income (₹ per ha) from Various Crops for Canal Lift Irrigators (Monsoon)

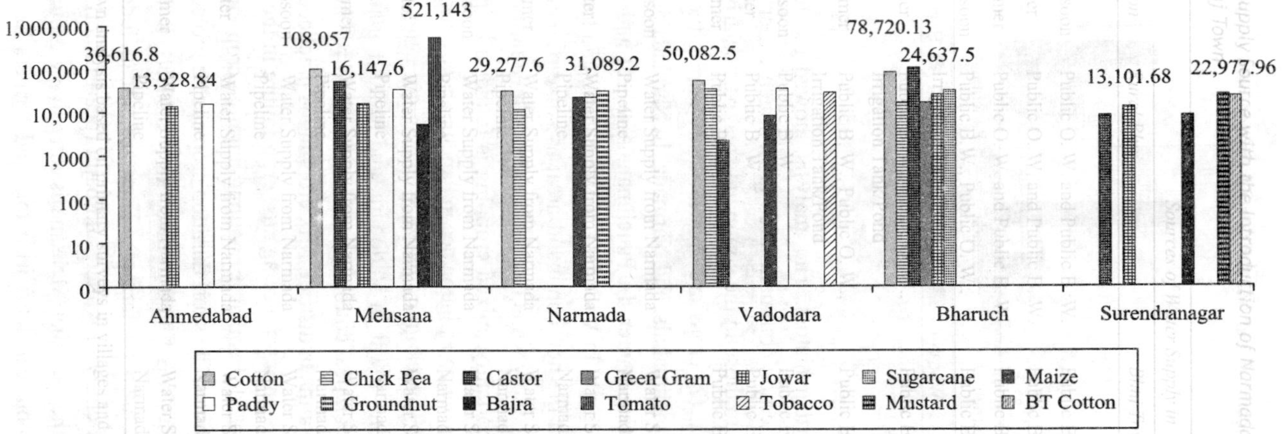

Source: Author.

Figure 6.9:
Current Net Income (₹ per ha) for Canal Irrigator (Winter and Summer)

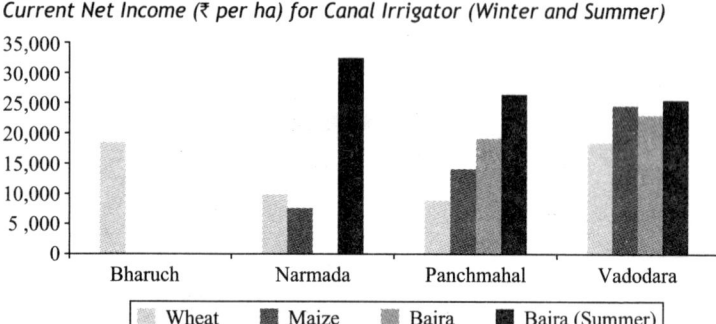

Source: Author.

with canal lift irrigators. The analysis showed the following: (1) the farmers in the non-command area grow much lesser number of crops as compared to those in the command area, particularly in Panchmahals and Narmada; (2) crops are grown mostly during the kharif season in Vadodara, Panchmahals, and Narmada. In Narmada district, only maize is grown during winter. In Panchmahals, wheat and maize are grown during that season. In Vadodara, maize, wheat, and tomatoes are grown by farmers in the non-command area, as some of them have access to water from shallow wells like the command area farmers.

More importantly, mixed cropping is found to be prevalent in these districts, with maize + pigeon pea in Panchmahals; cotton + maize, paddy + pigeon pea + black gram, maize + black gram + pigeon pea, and black gram + cotton+ jowar in Vadodara; and jowar + maize + pigeon pea, cotton + pigeon pea, cotton + maize and pigeon pea and maize+ paddy + cotton in Narmada district. Mix cropping is practiced in kharif season and is used as an insurance against crop failure in the wake of delayed monsoon. The idea is that if one long duration crop fails in the wake of delayed monsoon, the short duration crop can be protected.

Unlike what was found in these three districts, in Mehsana, several crops were grown by farmers. They are wheat, fodder crop, cumin, mustard, and chilly in winter; and bajra (pearl millet) in summer. This is owing to the fact that Mehsana is intensively groundwater irrigated region, and both canal lift irrigators and farmers having no access to

Figure 6.10:
Current Net Income (₹ per ha) from Various Crops for Canal Lift Irrigators (Winter and Summer)

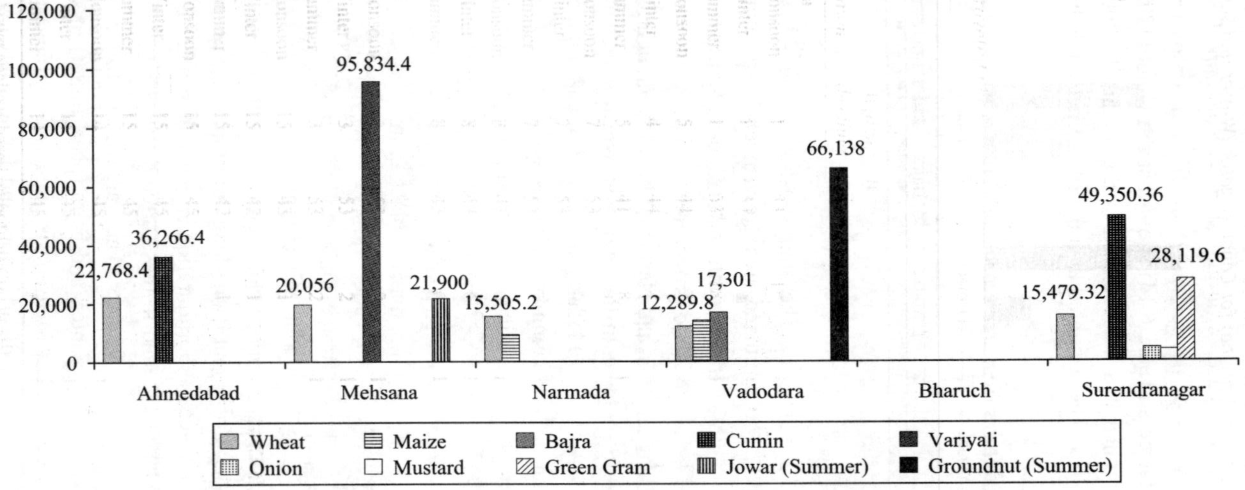

Source: Author.

canal water were able to get water from tube wells for irrigation either through water purchase, or through shared ownership of wells.

The estimates of net returns from different crops grown by non-command farmers (Table 6.14) show that the weighted average of net returns are much less as compared to that obtained by farmers in the command area. The weighted average of the net income per ha of the

Table 6.14:
Average Net Returns from Crop Production in Non-command Area

Name of Crop	Net Return from the Crop (₹ per ha) in			
	Mehsana	*Panchmahals*	*Vadodara*	*Narmada*
Kharif				
Cotton	26,698.0	9,596.6	14,502.6	14,694.4
Paddy		13,095.9	16,670.1	15,243.8
Chilly	21,065.6			
Cluster bean	33,369.4	98,754.5		
Pigeon pea		12,527.8	6,469.6	8,550.0
Mustard	26,306.3			
Castor	55,144.9		20,606.9	
Jowar	26,839.0			20,868.8
Bajra	19,971.3			
Sesame	9,621.1			
Choli	9,984.4			
Green Gram	6,750.0			
Black gram	3,988.6			
Maize		3,628.6	8,258.9	
Maize + Pigeon Pea		3,917.2	31,203.8	12,776.4
Cotton + maize			3,458.1	
Maize +black gram+ pigeon pea			7,878.6	
Maize + paddy +jowar				95,062.5
Paddy +pigeon pea + black gram			11,071.9	
Black gram + pigeon pea			33,970.3	
Cotton + pigeon pea			6,141.0	14,686.9
Jowar + maize + pigeon pea				34,272.3

(Table 6.14 Contd)

(Table 6.14 Contd)

Name of Crop	Net Return from the Crop (₹ per ha) in			
	Mehsana	*Panchmahals*	*Vadodara*	*Narmada*
Cotton + maize + pigeon pea				10,561.6
Cotton + paddy + jowar				
Pigeon pea + paddy + Cotton				30,425.0
Black gram + cotton + jowar			57,475.0	
Winter				
Wheat	18,129.0	15,669.3	8,771.2	
Chickudi grass	10,097.4			
Cumin	35,349.1			
Mustard	20,887.1			
Chilly	9,360.0			
Corn		9,275.4	7,338.3	3,680.8
Tomatoes			4,680.0	
Summer				
Bajra	29,756.3	18,591.0		
Groundnut			82,383.8	
Weighted average of the net return per ha of GCA	25,550.0	17,344.6	12,999.0	15,878.0
Average net return from crops for command area farmers per ha of GCA	118,009.0	34,678.0	55,171.0	33,146.0

Source: Authors' own analysis based on primary data.

gross cropped area for the command farmers ranged from ₹34,678 per ha in Panchmahals, ₹55,171 in Vadodara, to ₹33,146 in Narmada district. This is far higher than the average net returns per ha of the gross cropped area for their counterparts in the non-command area. For instance, in the case of Mehsana, the average net income per ha was estimated to be ₹25,550 against ₹118,009 per ha for the farmers who lift water from the canals. It is ₹17,345 per ha against ₹34,678 per ha in the command area for Panchmahals, ₹12,999 per ha against ₹55,171 per ha for Vadodara, and ₹15,878 per ha against ₹33,146 per ha for Narmada. One reason for the lower difference in the net income between command farmers and non-command farmers in the case of Panchmahals is the excessively high income farmers in the non-command are have earned from cluster bean. The market for cluster bean had gone up sharply during the past few months in the global market, resulting from the high demand for oil from its seed, and this had immensely benefited farmers in the study region also. The survey of non-command area farmers was undertaken more than a year after the survey in the command area was completed, and the market situation had changed during this time period.

Changes in Livestock Holding of Farmers

It was seen that in both canal command areas and areas now receiving irrigation through canal lift not only the gross cropped area has increased, but also the area under irrigated crops of kharif and winter has increased in all the locations. One of the outcomes of this is the increase in biomass availability throughout the year. Particularly, the increase in area under cereal crops such as wheat, bajra, and paddy would obviously increase the availability of leafy biomass (both dry and green biomass), the essential input for dairy production. This should ideally motivate the farmers to increase their livestock holding. Gujarat is famous for dairy development. In addition to what comes directly from the farms, increased availability of water running through canals in the rural landscapes is supporting the growth of natural vegetation, particularly grasses. Wild grasses grow abundantly on the banks of the small and large canals. Women moving around and cutting grasses along the banks of the canals is a common scene in the villages falling in the command area.

Though dairy output has been consistently increasing in the state over the past 40 years (Kumar et al., 2010b), in the recent past, there is no significant change in the livestock population from water-scarce regions which contribute to the milk output in a major way.[3] In fact,

[3] North Gujarat is one region which contributes disproportionately higher share to the state's dairy output and is characterized by some of the largest dairy processing units in Asia, including Dudh Sagar Dairy in Mehsana and Banas Dairy in Palanpur.

much of the addition in dairy output came from the replacement of low-yielding local breed of cows by high-yielding and crossbred cows and buffaloes.

But, Table 6.15 shows a phenomenon contrary to this happening in the villages of the command area. The livestock holding there has slightly increased over the past 5–6 years, the time frame considered for our analysis. The most significant increase is in Narmada district, where on average, one in every two farming households has increased the holding by one animal. This is followed by Panchmahals and Bharuch. In Vadodara, the net change in livestock holding is close to zero (refer Table 6.16). While this might appear quite negligible, what is notable is that the overall livestock holding of farmers is quite low here.

Similar trend was found in the case of farmers who are benefited by irrigation through canal lift, in Ahmedabad, Mehsana, Narmada, Vadodara, Bharuch, and Surendranagar (refer Table 6.16).

Such changes have impacted on the total income from dairy production in the villages as seen in Tables 6.17 and 6.18 for canal command areas and areas irrigated through canal lifting, respectively. In addition to the change in income affected by the increase in livestock holding, the average income from dairy animals has also increased, owing to the increase in milk yield and increase in the price of milk, resulting from the use of inputs such as dry and green fodder. Both the change in average income from dairy animals and the present average annual income are presented for farmers in canal command areas and farmers lifting water from canals, respectively, in Tables 6.17 and 6.18.

Incremental Farm Surplus

The values of incremental farm surplus with respect to different variables that were used to estimate the increase in farm surplus per ha of land of gravity irrigators and canal lift irrigators are provided in Tables 6.19 and 6.20, respectively. The final estimates of farm surplus are given in the last row for each location.

It can be seen from Table 6.19 that the increase in the average net income of farmers in different locations of Narmada command per ha of land is even higher than the average net income prior to Narmada

Table 6.15:
Change in Size of Animal Holding in Narmada Canal Command Areas per Farm Household

Location	Indigenous Cow			Crossbreed Cow			Buffalo			Bullock	Sheep
	In Milk	Dry	Calf	In Milk	Dry	Calf	In Milk	Dry	Calf		
Bharuch	0	0	0.07	0	0	0	0	0.02	0.07	0	0
Narmada	0.07	0.01	-0.01	0.01	0	0	0.18	0.04	0.11	0.04	0.05
Panchmahals	0.07	0	0	0	0.01	0	0.03	0.07	0	0	0
Vadodara	0.03	0.02	0.04	-0.01	-0.01	-0.02	0	-0.12	0.02	0.04	0

Source: Authors' own estimates based on data collected through a primary survey.

Table 6.16:
Change in the Size of Animal Holding in Areas Irrigated by Narmada Canal Lift

Location	Indigenous Cow			Crossbred Cow			Buffalo			Bullock	Sheep
	In Milk	Dry	Calf	In Milk	Dry	Calf	In Milk	Dry	Calf		
Ahmedabad	0.06	0.0	0.02	0.03	0.05	0.01	0.40	0.51	-0.01	0.0	0.0
Mehsana	0.06	0.0	0.02	0.0	0.0	0.0	0.23	-0.06	0.05	0.0	0.0
Narmada	0.10	0.06	0.0	0.0	0.0	0.0	0.19	-0.05	0.05	-0.15	0.0
Vadodara	-0.02	0.01	0.0	0.0	0.0	0.0	0.10	-0.02	0.00	0.15	0.0
Bharuch	-0.05	0.05	0.0	0.0	0.0	0.0	0.00	0.05	0.09	0.0	0.0
Surendranagar	0.00	0.12	0.06	0.0	0.0	0.0	-0.06	0.03	0.23	0.0	0.0

Source: Authors' own estimates based on data collected through a primary survey.

Table 6.17:
Annual Income (₹) of Dairy Farmers in the Canal Command Areas

Name of the Location	Indigenous Cow			Crossbred Cow			Buffalo		
	Input Cost	Gross Income	Net Income	Input Cost	Gross Income	Net Income	Input Cost	Gross Income	Net Income
Change in Annual Income from Livestock through Milk Production									
Bharuch	2,005.0	18,00.0	–205.0	0.0	0.0	0.0	1,480.0	12,135.0	10,475.0
Narmada	315.0	7,585.0	7,270.0	0.0	0.0	0.0	1,997.7	7,293.8	4,972.1
Panchmahals	6,000.0	15,080.0	13,080.0	0.0	0.0	0.0	1,417.0	9,289.9	7,015.3
Vadodara	4,460	5,665.6	5,435.0	930.0	9,760.0	6,600.0	1,843.4	11,099.5	9,313.9
Annual Income from Livestock through Milk Production after Narmada Waters									
Bharuch	4,685.0	9,900.0	5,215.0	0.0	0.0	0.0	9,540.0	22,195.0	12,475.0
Narmada	3,450.0	15,400.0	11,950.0	0.0	0.0	0.0	8,072.7	21,444.2	13,722.5
Panchmahals	6,000.0	15,080.0	13,080.0	0.0	0.0	0.0	8,276.7	20,310.5	12,033.7
Vadodara	8,460.0	11,668.1	7,438.1	7,620.0	25,960.0	18,340.0	7,996.7	22,966.0	15,255.0

Source: Authors' own estimates based on data collected through a primary survey.

Table 6.18:
Annual Income (₹) of Dairy Farmers Using Canal Water Using Lift

Name of the Location	Indigenous Cow			Crossbred Cow			Buffalo		
	Input Cost	*Gross Income*	*Net Income*	*Input Cost*	*Gross Income*	*Net Income*	*Input Cost*	*Gross Income*	*Net Income*
Change in Annual Income from Livestock through Milk Production									
Ahmedabad	3,373.0	11,217.6	10,680.6	−76.50	6,235.5	6,312.0	1,815.0	13,138.4	11,448.4
Mehsana	3,955.0	988.4	−381.6	0.0	0.0	0.0	1,805.5	11,256.5	9,553.8
Narmada	315.0	7,585.0	7,270.0	0.0	0.0	0.0	1,997.7	7,293.8	4,972.1
Vadodara	4,460	5,665.6	5,435.0	930.0	9,760.	6,600.0	1,843.4	11,099.5	9,313.9
Bharuch	2,005.0	1,800.0	−205.0	0.0	0.0	0.0	1,480.0	12,135.0	10,475.0
Annual Income from Livestock through Milk Production after Narmada Waters									
Ahmedabad	5,947.5	17,079	11,132.0	6,066.0	31,276.0	25,210.0	6,362.0	25,893.4	19,531.4
Mehsana	6,160.0	18,201.4	12,041.0	0.0	0.0	0.0	9,115.8	30,142.8	21,027.0
Narmada	11,340	20,900.0	9,560.0	0.0	0.0	0.0	9,170.0	29,694.2	20,524.0
Vadodara	7,062.0	9,405.0	2,343.0	0.0	0.0	0.0	6,792.8	19,544.2	12,751.5
Bharuch	900.0	7,950.0	7,050.0	0.0	0.0	0.0	7,030.0	20,271.0	13,241.0

Source: Authors' own estimates based on data collected through a primary survey.

Table 6.19:
Overall Farm Surplus of Canal Irrigators

Sr. No	Impact Variables	Farm Surplus Estimates for			
		Bharuch	*Narmada*	*Panchmahals*	*Vadodara*
1	Sample Size	29	60	30	60
2	Current Average GCA (ha per farmer)	5.16	0.85	6.02	5.64
3	Pre Narmada Average GCA (ha per farmer)	5.21	0.81	5.09	5.20
4	Average increase in GCA (ha per farmer)	−0.05	0.04	0.93	0.44
5	$\Delta A_{GROSS}/A_{GROSS}$	−0.010	0.047	0.154	0.078
6	Current Average NIA from canal (ha per farmer)*				
	(a) Kharif	5.08	0.75	4.24	4.10
	(b) Winter	0.08	0.09	1.35	1.49
	(c) Summer	0.0	0.00	0.43	0.05
7	Current Avg. GIA from canals (ha per farmer) (as per respondent survey)	5.09	0.72	0.98	1.01
8	NIA from SSP in the Region (ha)				
9	Average Net Income from Crops (pre-Narmada)(₹ per ha): ∅	6,885.4	5,484.8	13,649.0	12,707.9
10	Average Increase in Net Income-Crops (₹ per ha): ∅¹	47,043.2	27,661.5	21,029.2	42,463.2
11	Average Livestock Holding (Present) (no. per ha of GCA): N	0.14	1.68	0.20	0.52
12	Increase in Livestock Holding (no. per ha of GCA): ΔN	0.03	0.56	0.03	0.00
13	Average Net Income from Livestock (pre-Narmada) (₹ per Livestock): β	2,380.0	7,054.4	5,018.4	5,788.8

(Table 6.19 Contd)

(Table 6.19 Contd)

Sr.		Farm Surplus Estimates for			
No	Impact Variables	Bharuch	Narmada	Panchmahals	Vadodara
14	Increase in Average Net Income-Livestock (₹ per livestock): β^1	9,288.3	5,717.4	8,076.6	8,982.8
15	Farm Surplus per Ha of GCA (FARM$_{SURPLUS-UNIT}$)	**48,348.2**	**41,475.3**	**24,903.6**	**48,125.7**

Source: Authors' own analysis based on a primary survey.
*Estimates based on the data on irrigated crops in different seasons.

water introduction. Here, the average net incomes pre- and post-Narmada were worked out using the weighted average of the net income for different crops per unit area and the area under each crop, and, therefore, the change in average net income could have also resulted from a shift in the cropping pattern toward those which give higher return per unit area of land. That said, such a dramatic change has happened because of the slight change in the cropping pattern, shift from rain-fed to irrigated crops, introduction of high yielding varieties, and intensive farming using better inputs such as fertilizers and pesticides. This alone is sufficient to increase the farm surplus manifold without any changes in the gross cropped area. But, over and above this, there has been increase in income from dairy production as well, indicated by increase in the average net income from unit of livestock and increase in average number of livestock per unit of land.

Similar trend was found in the case of canal lift irrigators. The average increase in farm surplus for farmers in canal command from crops and livestock put together ranges from a lowest of ₹24,903 per ha for Panchmahals to the highest of ₹48,348 per ha for Bharuch.

Table 6.20 shows that the average increase in the net income from crop production post-Narmada is higher than the average net income from crops prior to Narmada. As a result, the net increase in farm surplus from crops and livestock was in the range of ₹14,195 per ha (for Narmada) to ₹109,855 per ha (for Mehsana). It is quite interesting to note that in spite of slight reduction in the gross area under production post-Narmada water introduction in Surendranagar, there has been a remarkable increase in the net farm surplus per ha in the order of ₹26,240 per ha.

Table 6.20:
Overall Farm Surplus of Farmers Irrigating through Canal Lift

Sr. No	Impact Variables	Farm Surplus Estimates for					
		Ahmedabad	Mehsana	Narmada	Vadodara	Bharuch	Surendranagar
1	Sample size	63	61	19	56	21	34
2	Current average GCA (ha per farmer)	17.28	10.79	8.20	7.46	5.80	5.23
3	Pre-Narmada average GCA (ha per farmer)	5.28	2.75	0.66	1.36	5.85	7.07
4	Average increase in GCA (ha per farmer)	12.00	8.04	7.54	6.10	−0.05	−1.84
5	$\Delta A_{GROSS}/A_{GROSS}$	0.69	0.74	0.92	0.82	−0.008	−0.35
6	Current average NIA from canal (ha per farmer)*						
	1) Kharif	3.25	1.90	0.62	1.29	5.80	3.58
	2) Winter	3.24	0.68	0.12	0.36		1.65
	3) Summer		0.01		0.01		0.0
7	Current average GIA from canals (ha per farmer) (as per survey)	14.17	8.46	2.89	5.68	24.91	3.84
8	NIA from SSP in the Region (ha)						
9	Average net income from crops (pre-Narmada) (₹ per ha): ∅	5,458.4	36,632.1	4,973.6	9,752.6	6,665.5	4,418.55

(Table 6.20 Contd)

(Table 6.20 Contd)

Sr. No	Impact Variables	Farm Surplus Estimates for					
		Ahmedabad	Mehsana	Narmada	Vadodara	Bharuch	Surendranagar
10	Average increase in the net income from crops (₹ per ha): $\varnothing^I$	11,751.4	81,377.0	8,846.3	13,493.9	49,524.9	23,508.1
11	Average livestock holding (Present) (no. per ha of GCA): N	0.15	0.10	0.11	0.28	0.14	0.26
12	Increase in livestock holding (no. per ha of GCA): ΔN	0.06	0.03	0.01	0.03	0.02	0.07
13	Average net income from livestock (pre-Narmada) (₹ per Livestock): β	7,793.0	11,491.1	11,609.2	1,998.9	3,426.2	30,157.0
14	Increase in the average net income from livestock (₹ per unit of livestock): β^I	11,018.0	8,375.0	5,998.8	8,565.3	10,475.0	8,400.0
15	Farm surplus per ha of GCA (FARM$_{\text{SURPLUS - UNIT}}$)	**17,662.2**	**109,855.0**	**14,195.5**	**23,926.8**	**51,002.5**	**26,248.70**

Source: Authors' own analysis based on a primary survey.
*As per estimates based on the data on crops which are irrigated.

Rises in Wage Employment of Farm Laborers

The labor absorption capacity of irrigated agriculture is well documented (Kumar and Amarasinghe, 2009; Bhattarai *et al.*, 2002; Narayanamoorthy and Deshpande, 2003). Irrigation can lead to increased rural employment and poverty reduction through several multiplier effects, depending on the structure of the rural economy (Bhattarai *et al.*, 2002), but farm labor remains the major contributor to wage employment and wage rates (Narayanamoorthy and Deshpande, 2003). Access to irrigation leads to expansion in the cropped area often with intensive cropping, with the result that the overall labor requirement for farming operations per unit area of holding increases. For crops such as paddy, the labor requirement is quite large for operations such as transplanting, weeding, and harvesting.

With a large expansion in the irrigated area over the past 6–7 years, the demand for agricultural labor has increased remarkably in south Gujarat districts of Vadodara, Bharuch, and Narmada and parts of Panchmahals. Table 6.21 shows the average number of days of wage employment obtained by farm laborers before and after the introduction of Narmada water in their villages, in four locations, and the percentage increase in the days of wage labor secured in each season owing to the increase in the irrigated crop production. Data are presented for male and female laborers, separately.

It can be seen that while during the summer months, the average wage employment (in number of days) obtained per wage laborer is lowest both pre- and post-Narmada, the increase in wage employment in agriculture post-Narmada is significant for summer months, except in Bharuch. During this season, a lot of demand for labor must be originating from the need for the preparation of farms for crop cultivation during kharif as the area cropped is very small in all the locations. As we have seen earlier, the land area under cropping is highest during the kharif season, followed by winter. Though no clear pattern seems to emerge from the results with regard to the increase in the number of days of wage employment, vis-à-vis either seasons or locations, the percentage increase was found to be the lowest in Bharuch district in all seasons and for both male and female workers. Probably, one reason for this is

Table 6.21:

Changes in Wage Employment in Agriculture Due to Canal Irrigation (Days)

Location	Male			Female		
	Monsoon	Winter	Summer	Monsoon	Winter	Summer
Before Narmada						
Panchmahals	36.9	49.7	28.1	38.0	48.5	23.8
Bharuch	55.7	76.6	59.0	55.9	77.6	59.5
Narmada	48.7	45.6	25.5	46.6	45.2	24.4
Vadodara	42.6	65.9	35.3	40.6	57.8	33.1
After Narmada						
Panchmahals	47.5	60.0	34.7	46.5	57.0	28.1
Bharuch	63.4	90.2	61.8	65.0	90.5	61.7
Narmada	58.5	63.0	36.0	57.9	60.5	34.7
Vadodara	50.4	78.6	44.4	47.2	72.0	36.3
Percentage Change						
Panchmahals	28.8	20.8	23.3	22.4	17.5	18.4
Bharuch	13.9	17.8	4.8	16.3	16.6	3.7
Narmada	20.1	38.3	41.1	24.3	33.9	42.3
Vadodara	18.5	19.2	25.8	16.3	24.6	9.9

Source: Authors' own analysis using data obtained from a primary survey.

that there was no expansion in the area under cropping with the introduction of Narmada water, though the area under irrigation increased. Again, Narmada district, which recorded the highest increase in the gross cropped area, showed the highest increment in wage employment in terms of the number of days (refer Figure 6.6).

Refer to Figure 6.11 for increase in wage employment in different seasons across locations in Narmada command.

Changes in Overall Socioeconomic Conditions of Farmers

Evidence worldwide shows that irrigation is reported to have positive impact on reducing poverty and improving the economic condition of farmers (Bhattarai *et al.,* 2002; Hussain and Hanjra, 2003; Narayanamoorthy and Deshpande, 2003). In India, the household consumption expenditure is being considered as an indicator of the economic

Figure 6.11:
Increase in Wage Employment (No. of Days) in Different Seasons, across Locations in Narmada Command

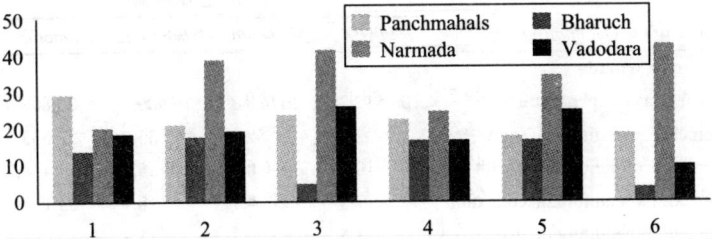

Source: Author.

condition of a family. Here, we have used the annual household expenditure for food, clothing, health and education and luxury and investment in savings as a proxy indicator of changing overall socioeconomic conditions of the farm households. Our analysis shows that on average, the total household expenditure has gone up remarkably for the sample farm households which received Narmada canal water for irrigation, post-Narmada. This cannot be interpreted as a major socioeconomic gain as the inflation will have its effect on the income and expenditure of the farm households. Hence, the best indicator is the percentage expenditure on basic food items. It is established from global evidence that with the increasing income (in real terms), families end up spending lesser proportion of their monthly income on food. Or in other words, richer families spend lesser proportion of their income for needs (food) than poor. It has also been found that with the increasing disposable income, families tend to spend more on social and religious ceremonies, travel and recreation, and other luxuries such as consumption of cigarettes and alcohol. Data were obtained for each one of these expenditure heads.

Our analysis of the pre- and post-Narmada scenarios clearly shows that the percentage expenditure for food items reduced considerably in all locations except Vadodara (Table 6.22). In Vadodara, the rise has been very marginal, that is, 0.30 percent. It is quite understandable in view of the fact that the rise in expenditure, which can be considered as a proxy for income, has not been very substantial in the case of Vadodara to offset the effect of inflation. As regards the proportion of the household expenditure on health and education, it dropped slightly in some

Table 6.22:
Impact of Canal Irrigation on Household Consumption Expenditure (₹ per Year)

Name of the Occupation	Bharuch	Narmada	Panchmahals	Vadodara
		Name of the Location		
Before Narmada				
Total family expenditure	109,517.2	36,628.3	70,909.8	70,403.4
Percent Expenditure on food	32.3	52.1	35.9	33.3
Percent Expenditure on health	10.1	8.9	5.3	4.3
Percent Expenditure on education	7.3	4.7	8.1	6.6
Percent Expenditure on luxury	18.5	26.2	15.4	21.4
Percent Savings in banks etc.	0.3	0.0	21.2	1.4
After Narmada				
Total family expenditure	163,058.6	54,284.2	162,389.4	97,311.0
Percent Expenditure on food	29.0	50.1	21.9	33.6
Percent Expenditure on health	9.4	9.3	4.1	5.4
Percent Expenditure on education	14.8	4.4	5.9	7.5
Percent Expenditure on luxury	22.3	27.2	10.3	19.3
Percent Savings in banks, etc.	0.4	0.0	12.3	0.1

Source: Author.
Note: Luxury includes travel and recreation, expenditure on liquor and cigarettes, and expenditure on social and religious ceremonies.

locations, while it increased in others. It is to be kept in mind here that the relationship between the real income and expenditure on education and health is not a straightforward one. Low levels of income rise will not lead to proportional increase in expenditure on education with the result that the percentage expenditure on this subsector of household finance would decline. Contrary to this, very high growth in family income can lead to disproportionately higher rise in expenditure on education, with the result that the percentage expenditure on the same can also go up. As regards health, it is quite likely that at low income levels, a substantial chunk of the family income goes in healthcare expenditure, whereas at higher levels of income, the percentage as well as aggregate expenditure on healthcare reduces (refer Figure 6.12).

A similar trend was found in the case of farmers lifting water form canals (Table 6.23). The annual family expenditure has soared in all the six locations. The largest increase in both aggregate and percentage

Figure 6.12:
Impact of Canal Irrigation on Family Expenditure

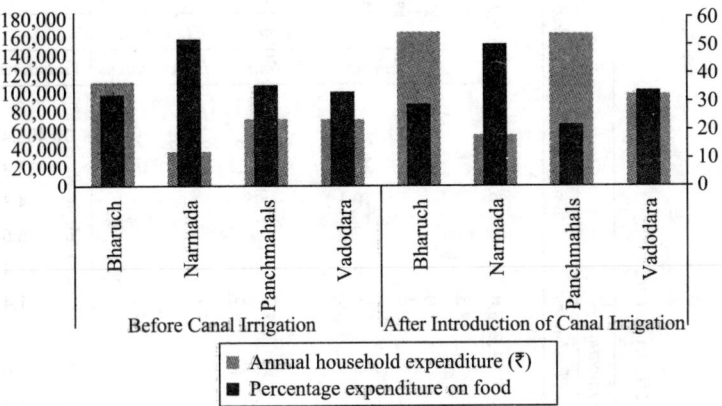

Source: Author.

terms was seen in Surendranagar district (from ₹65,526 to ₹201,000), followed by Bharuch (from ₹0.18 to ₹0.34 million), Mehsana (from ₹0.1 to ₹0.23 million), and Ahmedabad (from ₹0.12 to 0.19 million). Over and above, the percentage expenditure on food has dropped in five out of the five locations. The most significant difference is in the case of Surendranagar, followed by Mehsana, Vadodara, and Bharuch. In Surendranagar, it came down from58.8 percent to 40.4 percent. In Mehsana, it came down from 32.9 percent to 21.2 percent. In Vadodara, it came down from 38.5 percent to 29.2 percent of the annual household expenditure. In Bharuch, the expenditure on food dropped from 20.1 percent to 13.9 percent. In Ahmedabad, even prior to Narmada water introduction, the percentage expenditure on food was quite low (23.2), indicating prosperity, and it came down further to 21 percent. Only in the case of Narmada district, the percentage annual expenditure on food increased from 44.3 to 46.4. Increasing expenditure does not mean that the economic conditions have deteriorated, as the percentage expenditure is below 50 percent. Here, families spending greater percentage of their income for food may only mean that the quality of food intake has gone up considerably, with more protein and fat.

The impact was most visible in the villages of Limdi taluka of Surendranagar district through which the Saurashtra branch canal of the

Table 6.23:

Impact of Irrigation on Household Consumption Expenditure (₹ per Annum) of Canal Lift Irrigators

Name of the Occupation	Name of the Location						
	Ahmedabad	Mehsana	Narmada	Vadodara	Bharuch	Surendranagar	
Before Narmada							
Total family expenditure	119,890.9	102,309.1	42,447.1	71,276.0	176,098.8	65,526.3	
Percent Expenditure on food	23.2	32.9	44.3	38.5	20.1	58.8	
Percent Expenditure on health	2.9	3.0	6.0	5.0	5.9	11.8	
Percent Expenditure on education	6.3	7.9	6.7	9.9	7.5	12.7	
Percent Expenditure on luxury	14.0	22.3	36.0	16.9	13.4	14.0	
Percent Savings in banks etc.	6.8	0.0	0.0	0.0	5.7	2.7	
After Narmada							
Total family expenditure	190,520.0	229,124.8	60,160.8	123,794.2	337,755.2	201,008.9	
Percent Expenditure on food	21.1	21.2	46.4	29.2	13.9	40.4	
Percent Expenditure on health	4.5	3.0	7.5	5.7	4.9	4.8	
Percent Expenditure on education	7.4	9.6	5.8	9.8	7.9	12.2	
Percent Expenditure on luxury	15.7	8.8	32.1	14.0	13.5	41.2	
Percent Savings in banks etc.	8.7	1.7	0.0	0.0	5.9	1.5	

Note: Luxury includes travel and recreation, expenditure on liquor and cigarettes, and expenditure on social and religious ceremonies.

SSP is passing. The villages in this area are historically water deficit. The groundwater in the area is highly saline, making it even unfit for irrigation. The farmers used to take only rain-fed crops, mainly indigenous variety of cotton. The construction of the Saurashtra branch canal of the SSP came as a bounty for the farmers of this area, who are used to seeing only parched land after the rainy season, for generations. The enterprising farmers of this area had installed diesel engine sets to pump water from the branch canal into their fields. Water from the canal is conveyed through pipes which was often as long as 2–3 km. Many farmers who are not having adequate financial resources to purchase diesel pumps and distribution pipes have begun to purchase water from better-off farmers, who have pump sets. With canal water, they had been growing irrigated cotton, cumin, and wheat for the past three years. As a result, there was a major increase in the gross cropped area, unlike what was seen in the case of south and central Gujarat districts which are receiving Narmada canal water. The reason being that those areas were able to irrigate winter crops using the groundwater available from open wells and bore wells. The most significant change in their case was the replacement of rain-fed crops by their irrigated counterparts. With this, their profits from farming had gone up twofold.

Findings

- Area under irrigation has increased substantially in all the selected locations after the introduction of water by gravity through the Narmada canal system. Maximum increase in the average irrigated area through gravity irrigation was found to be in Bharuch—from 1.27 ha to 5.11 ha. In the case of canal lift irrigators, the impacts were striking. Maximum impact was found in Bharuch where the average area irrigated went up from 7.19 ha to 25.05 ha.
- With the introduction of water from the Narmada canal, farmers' dependence on wells and water purchase water has reduced. Well irrigation has become nonexistent in all the four selected locations which are receiving canal water by gravity. In the case of canal lift, the average area irrigated with purchased water went down

drastically in all the selected locations, with zero area in all locations except Bharuch. Similarly, well irrigation became nonexistent in all the locations except for Ahmedabad where it decreased from 6.8 ha to 0.05 ha post-Narmada. This phenomenon is quite comprehensible as many of these farmers were earlier dependent on well owners, who sell water at prohibitive prices. Whereas the well owners have to incur significant costs every year for the replacement and repair of pump sets and deepening of wells, better quality of water available from canals is another reason for switch over.

- The average gross cropped area per farmer increased after the introduction of water from Narmada canal, though the expansion was much more pronounced in the case of canal lift. For gravity irrigation, the highest increase in aggregate terms was in Narmada district, from 5.09 ha to 6.02 ha. For canal lift, the highest increase was for Ahmedabad—from 5.28 ha to 17.28 ha.

- After Narmada water, farmers shifted their cropping pattern toward remunerative crops. The gravity irrigators allocated greater proportion of the land to irrigated crops such as cotton (Bharuch), castor (Panchmahals and Vadodara), wheat, and maize in all study locations. Area under summer bajra too increased in three locations, viz., Narmada, Panchmahals, and Vadodara post-canal water. Similar changes in the cropping pattern were noticed among the farmers using canal lift. Overall, a notable increase in the area under kharif paddy, cotton, castor, chick pea, wheat, and maize was seen.

- There is a remarkable increase in the crop yield and net income returns per ha of land post-introduction of Narmada water for both gravity and canal lift irrigators for all crops. The yield improvement was quite considerable for castor, cotton, paddy, and wheat. This has been because of the following factors: (1) farmers providing irrigation to kharif crops, which were earlier grown under rain-fed conditions with the number of irrigated crops increasing in all study locations; (2) farmers growing longer duration, high-yielding varieties of the crops, viz., castor and cotton, which require irrigation, intensive use of labor, and fertilizers along with irrigation, manifested by increasing cost of cultivation.

- In both gravity and canal lift irrigated areas, the number of irrigated crops increased remarkably. The rain-fed crops were replaced by their irrigated counterparts. In gravity irrigated areas, the highest increase in the number of irrigated crops was found in Panchmahals (2–10), followed by Narmada (1–8) and Bharuch (1–7). In the case of canal lift, the highest was found in case of Vadodara (3–10). In none of the locations, farmers are now raising crops under rain-fed conditions. The fact that even prior to the Narmada Project, the farmers were growing crops in large areas without irrigation in the south Gujarat districts by virtue of the moderate–high rainfall and that these crops are now replaced by their irrigated counterparts partly explain the reason why there has not been much change in the gross cropped area under gravity.

- After Narmada canal irrigation, the net income from crop production increased remarkably across all study locations for both gravity irrigation and canal lift. For gravity irrigation, the increase in the net income was found to be the highest for cotton, across the locations, ranging from a lowest of ₹49,568 per ha in Panchmahals to a highest of ₹69,399 per ha in Vadodara. In the case of lift irrigators, rise in the net income was less than that of gravity irrigators, owing to the higher increase in the input cost for canal lifting. For cumin in Surendranagar, the net income per ha was ₹49,350 up from nil prior to Narmada water introduction.

- There has been a marginal increase in the livestock holding in the villages irrigated by gravity post-Narmada water introduction. The most significant rise was in Narmada district, where, on average, one in every two households added one animal. Similar trend was found in the case of farmers who are benefited by irrigation through canal lift, in Ahmedabad, Mehsana, Narmada, Vadodara, Bharuch, and Surendranagar.

- There were changes in the average annual income from dairy production in the villages for both canal command areas and areas irrigated through canal lifting. For gravity irrigation, the highest increase in the average annual income from dairy was found in villages of Bharuch—from ₹10,475 to ₹17,585—and for canal lift in the villages of Narmada district—from ₹4,972 to ₹20,524 per animal unit.

- There was an increased demand for farm labor especially in south Gujarat districts of Vadodara, Bharuch, and Narmada and parts of Panchmahals due to large expansion in irrigated area owing to the availability of canal water. The increase in wage employment in agriculture post-Narmada is significant for summer months, except in Bharuch.

- Analysis of changes in farm surplus shows that the average annual incremental net income from crop production per ha of the gross cropped area owing to gravity irrigation from Narmada varied from a lowest of ₹21,029 for Panchmahals to ₹47,043 for Bharuch. In the case of canal lift, the incremental net income varied from the lowest of ₹13,494 for Vadodara to the highest ₹81,377 for Mehsana. The incremental net income per annum from dairying varied from ₹5,717 per livestock unit for Narmada to ₹9,288 per livestock unit for Bharuch under gravity irrigators. In the case of canal lift, it varied from ₹5,999 for Narmada to ₹11,018 for Ahmedabad.

- The sharp economic impact of canal irrigation on farming is corroborated by the fact that the overall net income from crop production per ha of the gross cropped area, based on weighted average of the net income per ha of crop for different crops, was found to be much higher for farmers receiving canal water either through gravity or through lift as compared to the non-command farmers in all the four locations surveyed. The net income returns per ha of gross cropped area (GCA) of crop for the non-command area farmers were ₹25,550 in Mehsana, ₹17,344 in Panchmahals, ₹12,999 in Vadodara, and ₹15,878 in Narmada.

- The increase in the overall farm surplus per ha of the gross cropped area in gravity irrigated area was estimated to be varying from a lowest of ₹24,903 per ha per annum in Panchmahals to ₹48,348 per ha in Bharuch. In Vadodara, it was ₹48,126 per ha, and in Narmada, it was ₹41,475 per ha. This is an enormous gain, when we consider the fact that the areas which now receive water from Narmada canals by gravity mostly fall in the high rainfall areas.

- As regards the farmers engaged in canal lifting, the increase in overall farm surplus per ha of the gross cropped area was estimated to be varying from a lowest of ₹14,195 per ha in Narmada to the highest of ₹109,855 per ha in Mehsana. The values were ₹17,662 per ha in Ahmedabad; ₹23,927 per ha in Vadodara, and ₹51,002 per ha in Bharuch. The exceptionally high income gain in the case of Mehsana is because of the high-valued crops such as fennel being grown by farmers in that area.

- Total household expenditure for the farm households which received Narmada canal water for irrigation has remarkably increased. More importantly, the proportion of the expenditure for food items reduced considerably in all locations, a strong indication of the poverty-reduction impact of canal irrigation. Similar trend was found observed in the case of areas benefited by canal lift. The largest increase in annual family expenditure in both aggregate and percentage terms was seen in Bharuch—from ₹0.18 million to ₹0.34 million. The percentage expenditure on food has dropped in five out of the six locations. The expenditure on children's education was also found to have gone up. The overall income of rural families, particularly farmers and wage laborers, had also gone up post-Narmada.

***

The direct economic impacts of irrigation from the SSP—increased income from crop and milk production, better living standards for the farm households and increased demand for wage labor in agriculture—are remarkable. But such big gains have not come from the increase in cropping intensity, but from the replacement of rain-fed varieties by their irrigated counterparts. But, with intensive well irrigation already happening in many parts of the design command, except the tribal areas of Bharuch, Narmada, and Panchmahals, the region covered by the SSP irrigation is agriculturally far more developed than some of the countries in Sub-Saharan Africa.

Recently, Grey and Sadoff (2007) argued on the basis of illustrations from many parts of the world that the returns on investment in water development would be high in regions which are in the initial stages of resource development. Kumar *et al.* (2008) argued, on the basis of the analysis of datasets from 22 countries around the world that for countries falling in semiarid and arid tropics, large water storage would be a prime mover for reducing hunger and fostering economic growth. There are many developing countries particularly in Sub-Saharan Africa where the country-average of per capita water storage is abysmally low. These are also tropical countries with semiarid to arid climatic conditions. A fraction of the arable land in these countries is cultivated, and a smaller fraction irrigated. These countries are not only economically poor, but have very low social development indicators. Problems of chronic hunger and poverty and inadequate access to water for basic survival are characteristic of these countries (Kumar *et al.*, 2008). Building large water storages would be a prerequisite for improving water security in these countries (Kumar *et al.*, 2008; Shah and Kumar, 2008; Priscoli and Briscoe, 2011), which can come from expansion in cropping intensity as well as irrigation intensity. However, this has to be accompanied by the right kind of institutional and policy framework to ensure sustainable water use (Kumar *et al.*, 2008; Shah and Kumar, 2008).

7

Drinking Water Supplies from Narmada: Socioeconomic Impacts

Historically, Gujarat has acute shortage of drinking water in many regions, which are poorly endowed with fresh water resources. Problems worsened during droughts (Raj, 1991). The depletion of groundwater and deterioration of groundwater quality had increased in magnitude and spread (Kumar *et al.*, 2000; Das and Gupta, 2010). There have been substantial efforts since the late 1990s by several state agencies[1] to enhance the coverage of public water supply schemes, and improve the quantity, quality, and reliability. One of the major problems which the state government tried to address is the unequal distribution of water resources across space, and the inequitable allocation of the limited water resources in poorly endowed regions to agriculture, based on the basic needs of the population. For addressing these problems, techno-institutional solutions were sought, such as bulk transfer of reliable water supplies from Narmada to poorly endowed regions such as Kutch and Saurashtra, with increased allocation during summer months (Talati and Kumar, 2005). This can be seen from a remarkable increase in the number of habitations covered (see Table 7.3, as cited in Gujarat State Drinking Water Infrastructure Co. Ltd [2000]). According to the official statistics, there are only 1,271 habitations existing which are not covered fully by proper water supply (that is, 40 lpcd).

[1] These agencies include Gujarat Water Supply and Sewerage Board (GWSSB), Water and Sanitation Management Organization and SSNNL.

A study in 2005 had shown remarkable improvement in the water supply in Saurashtra and Kutch villages after the introduction of water from Narmada and Mahi through long-distance pipelines (citizen's monitoring of water supply based on Narmada, IWMI-TATA-Project [ITP]), though water supply was not very regular in many villages. Since 2005, substantial progress is made by GWSSB in implementing the Narmada-based pipeline scheme in the two regions of Saurashtra and Kutch, with thousands of villages added to the scheme. As on October 2010, the water supply grid already started covering 10,528 of the 18,066 villages in the state. Of these, 6,513 villages are covered by Narmada canal-based supply alone, which covers a length of 1,957 km. The Narmada canal-based water supply is expected to reach a total of 9,633 villages in the state.[2]

One of the biggest criticisms which the SSP received during the initial years of its planning was that water would never reach the parched areas of drought-prone Saurashtra and Kutch to ensure drinking water security to that region, a major argument used by the proponents of the SSP to justify the large investment in the project. The anti-dam activists argued that drinking water security was being used by the state officials engaged in planning the SSP as a major ploy to justify a project, which would supply irrigation water to an already prosperous region of south and central Gujarat. Hence, it is important to undertake an assessment of the impact of water supply provision on the communities in the region.

The Current Drinking Water Scenario

The importance of water supply based on Narmada canal lies in its dependable nature. The local groundwater-based sources which communities in Saurashtra used to depend on for drinking water supplies, failed during drought years. During the historic drought of 1985–1987, water had to be taken to Rajkot city of Saurashtra, through rail wagons, the cost of which was more than that of desalination of seawater.

Given the fact that only a small fraction of the total volumetric allocation from the Narmada basin to the state of Gujarat (9.0 MAF or

---

[2] http://www.environmentportal.in/files/SSP_Report.pdf

10,800 MCM per annum) is required for meeting the drinking water supply needs of nearly 10,000 villages and 125 towns proposed to be covered under the scheme, and flows in Narmada are far more dependable than the flows in the rivers and renewable recharge in the aquifers of the regions under consideration (such as Saurashtra and Kutch), even during years of worst drought, water supply from Narmada canal-based schemes would not get affected. On the contrary, during drought years, the hard rock aquifers in Saurashtra and Kutch would be completely depleted, drying up the local drinking water supply sources. This really enhances the value of water available for domestic consumption from Narmada. As Alagh *et al.* (1995) noted, as alternative sources of water would not be available in the region during droughts, the value of the water supplied from Narmada should be treated as equal to the cost of desalination to produce and supply water.

The availability of dependable water supplies throughout the year had liberated thousands of villages from their dependence on tanker water supply, with the state as well as communities saving substantial amount of resources spent on water transportation (Modi, 2010). See Table 7.1.

Narmada Drinking Water: Socioeconomic Impact

One of the important landmarks in the implementation of the SSP is that it could carry water to far off places like Kutch, which remained thirsty for nearly two centuries, through the Narmada main canal and long

Table 7.1:
Progress in Drinking Water Supply Coverage in Gujarat Villages over Time

Year	Fully Covered	Partially Covered	Not Covered	PC + NC
1993	19,970	8,981	1,318	10,299
1995	21,994	6,553	1,722	8,275
1999	25,193	4,639	437	5,076
2000	25,414	3,426	1,429	4,855
2002	28,396	1,777	96	1,873
2003	29,027	1,213	29	1,242

Source: Gujarat State Drinking Water Infrastructure Co. Ltd, 2000.

pipelines. Historically, water had played a significant role in shaping the identity of Kutch. Villagers across the region believe that the lack of water was the cause of large out-migration. Water scarcity is attributed to low rainfalls, high aridity, and frequent droughts (Mehta, 1997; IRMA/UNICEF, 2001; Kumar, 2002). Memories are often evoked of the times when Kutch was green and verdant, when one of the tributaries of river Indus flowed through the region and watered the Banni grassland (Kutch Development Forum, 1993).

The cherished longing of the people of Kutch to get water from outside the district was the outcome of the injustice that it had to face vis-à-vis water due to natural and political circumstances. An earthquake in 1812 changed the course of the tributary which was flowing into Kutch. After the partition in 1947, traditional sociocultural links with the Sind region were severed and plans to divert Indus waters to Kutch were abandoned. Riparian agreements over the distribution of water from the rivers of Punjab also left out Kutch. Kutch was then given the assurance that water from the Rajasthan canal (the Indira Gandhi Nehar Project) would be diverted to the area (Thakker, 1988, as cited in Mehta, 1997). But, such hopes were belied.

These developments led the people of Kutch to feel bitter as they felt that the region has been successively marginalized. There is a widespread belief in Kutch that due to the harsh climate, erratic water supply, declining groundwater resources, and frequent droughts, the only solution to the problem if desiccation is to get water from the rivers of Gujarat (Kutch Development Forum, 1993). Thus, all hopes are pinned on the SSP (Mehta, 1997).

The real impact of a water supply intervention in terms of its ability to change the household dynamic could be assessed by examining the degree of dependence of the communities on the new source after it becomes fully operational. At the next level, the impact could be assessed by examining the changes in volumetric water use. Our analysis for rural Jamnagar shows that after the introduction of Narmada water, the community in rural areas started using only the water from Narmada pipeline system for drinking and cooking and domestic uses, whereas they used to depend on public bore wells and common stand posts prior to the introduction of Narmada water system (Table 7.2). People have to walk long distances to fetch water from distant sources such as farm

Table 7.2:
Changing Source of Water Supply with the Introduction of Water from Narmada Pipeline: Rural Jamnagar and Jamnagar Town

Type of Water Use	Name of Season	Names of Water Sources Tapped to Meet the Needs	
		Rural Jamnagar	Jamnagar Town
Before Narmada			
Drinking and cooking	Monsoon	CSP, Public B. W	Tap Water Supply, Public Bore Well, Other Sources
	Winter	CSP, Public B. W	Tap Water Supply, Public Bore Well, Other Sources
	Summer	CSP, Public B. W	Tap Water Supply, Public Bore Well, Other Sources
All other domestic uses	Monsoon	CSP, Public B. W	Tap Water Supply, Public Bore Well, Other Sources
	Winter	CSP, Public B. W	Tap Water Supply, Public Bore Well, Other Sources
	Summer	CSP, Public B. W	Tap Water Supply, Public Bore Well, Other Sources
Livestock	Monsoon	Own O.W, Public B. W	
	Winter	Own O. W, Public B. W	
	Summer	Own O. W, Public B. W	
After Narmada			
Drinking and cooking	Monsoon	Water Supply from Narmada Pipeline	Water Supply from Narmada Pipeline
	Winter	Water Supply from Narmada Pipeline	Water Supply from Narmada Pipeline
	Summer	Water Supply from Narmada Pipeline	Water Supply from Narmada Pipeline
All other domestic uses	Monsoon	Water Supply from Narmada Pipeline	Water Supply from Narmada Pipeline
	Winter	Water Supply from Narmada Pipeline	Water Supply from Narmada Pipeline
	Summer	Water Supply from Narmada Pipeline	Water Supply from Narmada Pipeline
Livestock	Monsoon	Water Supply from Narmada Pipeline	
	Winter	Water Supply from Narmada Pipeline	
	Summer	Water Supply from Narmada Pipeline	

Source: Author.

wells during months of summer, when most of the groundwater-based public sources in the village dry up. Whereas in Jamnagar town, while the communities used tap water supply from public bore wells and water from other sources including the desalinated water supplied from Reliance Petrochemicals prior to Narmada water introduction, they have now totally shifted to Narmada pipeline-based water supply. It is to be recalled here that Jamnagar town used to face one of the most severe water shortage in entire Gujarat.

Similar data were collected through primary survey and compiled for rural Bhuj and Bhuj town (Table 7.3). Here again, the rural communities now fully depend on a single source, that is, Narmada pipeline, whereas they used to depend on at least three different sources of water, including public open wells and bore wells and irrigation tanks. Further, they used to depend on public open wells and bore wells for drinking and cooking. For other domestic uses, different sources, including the irrigation tanks/ponds, were tapped. For livestock drinking, the public bore wells were used. Such selection is always made with due consideration to the match between the quality of water from these sources and water quality requirements for the type of use. Tapping distant sources also meant that people travel long distances to fetch water spending enormous time. Unlike in the case of Jamnagar, in Bhuj, even urban communities are engaged in livestock keeping.

Changes in Frequency of Water Supply

It is well established that the distance traveled and the time taken to procure water impact on the amount of water collected and daily water consumed by households (WHO/UNICEF, 2000; Howard and Bartram, 2003). Hence, the frequency of water supply would significantly influence the daily water consumption by households as it determines the actual number of trips to be performed by the water collector. Data shows that there is lack of any significant improvement in the frequency of water supply post-Narmada, in spite of major improvement in water availability situation in terms of dependability. On the contrary, in both rural and urban areas of Bhuj, the frequency of water supply actually reduced after the introduction of Narmada canal-based water supply.

Table 7.3:
Changes in Water Supply Source with the Introduction of Narmada Water in Rural Bhuj and Bhuj Town

Type of Water Use	Season	Sources of Water Supply in	
		Rural Bhuj	*Bhuj Town*
Before Narmada			
Drinking and cooking	Monsoon	Public O. W. and Public B. W.	Public Bore Well
	Winter	Public O. W. and Public B. W.	Public Bore Well
	Summer	Public O. W. and Public B. W.	Public Bore Well
All other domestic uses	Monsoon	Public B.W., Public O. W., Irrigation Tank/Pond	Public Bore Well
	Winter	Public B. W., Public O. W., Irrigation Tank/Pond	Public Bore Well
	Summer	Public B. W., Public O. W., Irrigation Tank/Pond	Public Bore Well
Livestock	Monsoon	Public B. W.	Public Bore Well
	Winter	Public B. W.	Public Bore Well
	Summer	Public B. W.	Public Bore Well
After Narmada			
Drinking and cooking	Monsoon	Water Supply from Narmada Pipeline	Water Supply from Narmada Pipeline
	Winter	Water Supply from Narmada Pipeline	Water Supply from Narmada Pipeline
	Summer	Water Supply from Narmada Pipeline	Water Supply from Narmada Pipeline
All other domestic uses	Monsoon	Water Supply from Narmada Pipeline	Water Supply from Narmada Pipeline
	Winter	Water Supply from Narmada Pipeline	Water Supply from Narmada Pipeline
	Summer	Water Supply from Narmada Pipeline	Water Supply from Narmada Pipeline
Livestock	Monsoon	Water Supply from Narmada Pipeline	Water Supply from Narmada Pipeline
	Winter	Water Supply from Narmada Pipeline	Water Supply from Narmada Pipeline
	Summer	Water Supply from Narmada Pipeline	Water Supply from Narmada Pipeline

Source: Author's own analysis based on primary surveys in villages and towns.

Nearly 50 percent of the households in rural and urban Bhuj reported as receiving water supply once in 2–3 days post-Narmada, whereas the percentage of people getting water supplies at that frequency was only nearly 15 (that is, 9 and 11 households, respectively).

At low frequencies of water supply, as Arghyam/IRAP (2010) has shown, the household's ability to collect and use adequate quantities of water for daily needs depends heavily on the availability of storage systems, even if the distance that need to be covered or the time required to fetch water do not change. A large proportion of the families (ranging from 25 percent for rural Bhuj and rural Jamnagar to 75 percent for Bhuj town to around 80 percent for Jamnagar town) received water supply only on alternate days in the four locations (see in Table 7.4), signifying the importance of intermediate storage systems.

Changes in Distance to the Source of Water Supply

The effort required by people to fetch water is an important parameter in assessing the performance of water supply schemes (Howard and

Table 7.4:
Frequency of Water Supply before and after Narmada Canal-based Piped Water Supply

	No. of HHs Having Access to Water Supplies						
Location	*Daily*	*Alternate day*	*Once in 2–3 days*	*Once in 4 days*	*Once in 5 days*	*Once in week*	*Twice a day*
Before Narmada							
Bhuj rural	8	16	9	11	–	15	–
Jamnagar rural	42	16	–	–	–	–	–
Bhuj urban	–	47	11	2	1	–	–
Jamnagar urban	17	42	3	–	–	–	–
After Narmada							
Bhuj rural	4	3	29	11	–	13	1
Jamnagar rural	42	16	–	–	–	–	–
Bhuj urban	–	22	31	7	1	–	–
Jamnagar urban	17	42	3	–	–	–	–

Source: Authors' own analysis based on primary data collected from rural and urban households in Bhuj and Jamnagar.

Bartram, 2003; WELL, 1998; WHO/UNICEF, 2000). Evidences world-wide show that volume of water used in a home is sensitive to gross differences in service levels in water supply, which include the travel distance or the return time taken to procure water (WHO/UNICEF, 2000).

Table 7.5 shows the percentage of households who get water within the dwelling, or within the dwelling premises, within 0.2 km distance

Table 7.5:
Total Distance Traveled per Day for Collection of Water (m)

		Percentage Households Who Get Water					
Location	*Season*	*Within dwelling*	*Within premise*	*Within 0.2 km*	*Within 0.2–0.5 km*	*Within 0.5–1.0 km*	*More than 1 km*
Before Narmada							
Rural Bhuj	Monsoon	1	33	18	3	–	4
	Winter	1	33	18	3	–	4
	Summer	1	33	18	3	–	4
Rural Jamnagar	Monsoon	5	44	7	–	3	–
	Winter	4	44	7	–	3	
	Summer	5	41	8	1	3	
Bhuj town	Monsoon	7	53				
	Winter	7	53				
	Summer	7	53				
Jamnagar town	Monsoon	8	43	6	1		4
	Winter	8	43	6	1		4
	Summer	8	43	6	1		4
After Narmada							
Rural Bhuj	Monsoon	3	53	2	1		
	Winter	3	53	2	1		
	Summer	3	53	2	1		
Rural Jamnagar	Monsoon	15	43	1			
	Winter	15	42	1			
	Summer	15	42	1			
Bhuj town	Monsoon	15	45				
	Winter	15	45				
	Summer	15	45				
Jamnagar town	Monsoon	14	45	2	1		
	Winter	14	45	2	1		
	Summer	14	45	2	1		

Source: Authors' own analysis based on primary survey data.

of the dwelling, etc., both before and after the introduction of Narmada canal-based water supply. It clearly shows that a greater proportion of the sample households get water within the dwelling or within the premises after the introduction of Narmada. For instance, while nearly 33 percent of the sample HHs in rural Bhuj got water in within their dwelling premises, the figure went up to 53 percent after Narmada water supplies. Similarly, while only 4–5 percent of the sample HHs in rural Jamnagar got water in their dwelling, the figure went up to 15 percent post-Narmada. Similar trend was noticed in case of Bhuj and Jamnagar towns. As regards the proportion of population depending on distant sources, the figure went down, indicating that many of these HHs are getting water supplies close to their dwelling.

Changes in the Time Spent on Water Collection

We have already seen that with the introduction of Narmada waters, communities in both rural and urban settings have better physical access to water supplies. Hence, this is naturally reflected in the time spent on water collection. Analysis of relevant data collected from the sample households are presented in Table 7.6. It shows that there has been saving in the time spent on water collection (Table 7.6). During monsoon, the time saving was to the tune of 6 and 19 minutes in rural and urban Bhuj, respectively. However, rural areas reported more time savings in comparison to the urban areas in Jamnagar. Similar trend in time saving on water collection was found during winter season. There is also increase in the time saving during summer season, particularly in the case of rural Bhuj and Jamnagar towns, when the magnitude and extent of water shortage was on high prior to Narmada water introduction.

In spite of the significant reduction in the distance travelled, the reduction in time spent on water collection is only marginal. One reason for this could be the reduction in the frequency of water supply encountered by a significant proportion of the households, in rural Bhuj and Bhuj town, which meant that the community members have to fetch more water at a time, indicating more trips to the source for collecting water.

Table 7.6:
Impact of Water Supply on the Average Time Spent on Collecting Water for Domestic Uses in Different Locations

Location	Average Time Spent (Minutes)		
	Monsoon	*Winter*	*Summer*
Before Narmada			
Bhuj rural	95.0	102.0	102.0
Jamnagar rural	69.8	69.8	69.8
Bhuj urban	38.0	38.0	38.0
Jamnagar urban	58.0	58.0	58.0
After Narmada			
Bhuj rural	89.0	90.0	90.0
Jamnagar rural	55.0	55.0	55.0
Bhuj urban	31.0	31.0	31.0
Jamnagar urban	54.0	54.0	52.0
Time Saved			
Bhuj rural	6.0	12.0	12.0
Jamnagar rural	21.1	21.0	21.0
Bhuj urban	19.0	19.0	19.0
Jamnagar urban	8.0	8.0	11.0

Source: Author.

Domestic Water Consumption by the Households

The amount of water consumed by the households is as important a factor in deciding the health risks associated with water supply as the quality of water supplied and the hygiene practices (Esrey *et al.*, 1985, 1991). Our analysis of monthly household level water usage, including livestock water use, does not show any significant change in terms of the volumetric daily water consumption by the household post-Narmada canal-based piped water supply. But, as previously pointed, the frequency of water supply, which has reduced in two locations, might have had its own impact on household water consumption owing to the increase in water storage requirements. Apart this, the overall impact of "reduction in distance to the source" on "time required for water collection" remained marginal because of the reduced frequency.

Even at present, the average time taken for the collection of water is quite significant—above 30 minutes—in all the four locations. As research-based evidence available shows a clear relationship between travel time and water consumption (WELL, 1998), the real impact of reduction in time spent on procuring water on water consumption by the communities become evident when the time taken becomes less than five minutes. The impact of reduction in time spent on water collection on water consumption is expected to be almost nil in the time range of 5–30 minutes (Cairncross and Feachem, 1993), becomes explicit afterwards, with the water consumption likely to reduce in rural areas to a bare minimum where only the basic consumption needs can be met (Aiga and Umenai, 2002).

This essentially means that the water use would have substantially increased with the introduction of Narmada pipeline water supply, if the frequency of water supply remained at least the same. But, this no way indicates that the water supply situation is poor today. In areas such as Bhuj and Jamnagar, which used to face adverse problems of water quality, the performance of water supply cannot be assessed in relation to the quantum of water used per day, particularly when the households have not expressed any dissatisfaction with the quantum of water supplied. Instead, the quality of the supplied and the changes in water quality after the introduction of a new source are the aspects to be examined.

It is important to know the current level of consumption by the households and how far this is adequate for reducing the health risks. The estimates of daily water consumption in different locations, pre- and post-Narmada are presented in Figure 7.1. It shows that the levels of consumption are quite adequate today, with the per capita consumption exceeding 50 liter per day with very low health risk implications, though the aggregate quantity of water consumed has not changed. Going by the current level of the use of water, the access can be termed as "intermediate" (Howard and Bartram, 2003). Here, sufficient water is available to meet the domestic water needs, with sufficient quality, and therefore, such communities can be said to have effective household water security (Van der Hoek et al., 2002).

The estimates of seasonal domestic water consumption by the sample households in different subsectors, derived from the primary survey data,

Figure 7.1:
Per Capita Daily Domestic Water Use (Liters)

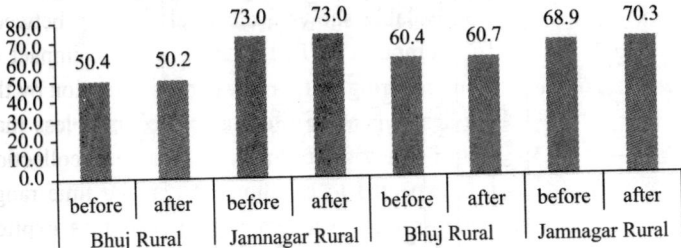

Source: Author.

are presented in Tables 7.7, 7.8, and 7.9, respectively, for rainy season, winter, and summer. As could be seen from the tables, there is no notable difference in the seasonal water use of the family in any of the subsectors of domestic water economy, except minor differences in water use for drinking and cooking. The difference was only 13 liter during rainy season and 8 liter during summer in rural Bhuj; the difference was highest, that is, 161 liter during rainy season and winter season in rural Jamnagar, but only 65 liter in summer there. In Bhuj town also, there was some improvement in water use for drinking and cooking, to the tune of 118 liter each during rainy season and winter season, and 151 liter during summer season. But, even this difference is quite noteworthy, in view of the fact that consumption would require more water if it contains salts, particularly in hot and arid climates. Several families have reported that since the water supplies from Narmada canal-based scheme are free from salts (TDS), they actually required much less water to quench the thirst, which was also evident from a past survey as reported by Talati and Kumar (2005).

Besides drinking and cooking, another subsector where some improvement was seen is family's water use in toilets in Jamnagar town. It went up from 9,997 to 11,603 liter in summer. Since the same trend was not seen for other seasons, one needs to exercise some caution while drawing inference on the basis of this data. What is important to remember is that the absolute figures are not so important here, but the trend, if any, is.

Table 7.7:
Seasonal Domestic Water Consumption by the Sample Households: Rainy Season

	Volumetric Water Consumption during Rainy Season: Pre- and Post-Narmada								
	Bhuj Rural		Jamnagar Rural		Bhuj Urban		Jamnagar Urban	
Type of Domestic Water Use	Before	After	Before	After	Before	After	Before	After
Drinking and cooking	2,130	2,143	2,059	2,220	1,919	2,037	2,266	2,256
Washing	6,461	6,495	9,223	9,223	6,993	7,052	8,750	8,750
Bathing	9,839	9,839	17,354	17,354	12,413	12,472	17,840	17,840
Toilet use	6,424	6,424	9,435	9,435	8,336	8,336	9,997	9,997
Cleaning utensils	3,437	3,468	5,937	5,937	4,593	4,593	5,526	5,526
Scooter/car washing	0.00	0.00	0.00	0.00	383	383	413	413
Gardening	0.00	0.00	0.00	0.00	750	750	–	–
Floor cleaning	2,712	2,712	3,727	3,727	3,433	3,433	3,667	3,667

Source: Authors' own analysis based on data collected through a primary survey in four locations.

Table 7.8:
Seasonal Domestic Water Consumption by the Sample Households: Winter

| | Volumetric Water Consumption during Winter Season: Pre- and Post-Narmada | | | | | | | |
| | Bhuj Rural | | Jamnagar Rural | | Bhuj Urban | | Jamnagar Urban | |
Type of Domestic Water Use	Before	After	Before	After	Before	After	Before	After
Drinking and cooking	2,130	2,138	2,059	2,220	1,919	2,037	2,265	2,255
Washing	6,461	6,495	9,223	9,223	6,861	6,964	8,750	8,750
Bathing	9,839	9,839	17,354	17,354	12,413	12,472	17,840	17,840
Toilet use	6,424	6,424	9,435	9,435	8,336	8,203	9,997	9,997
Cleaning utensils	3,437	3,468	5,937	5,937	4,561	4,593	5,526	5,526
Scooter/car washing	0.00	0.00	0.00	0.00	383	383	413	413
Gardening	0.00	0.00	0.00	0.00	750	750	–	–
Floor cleaning	2,712	2,712	3,727	3,727	3,433	3,433	3,667	3,667

Source: Authors' own analysis based on data collected through a primary survey in four locations.

Table 7.9:
Seasonal Water Consumption by the Sample Households: Summer

	Volumetric Water Consumption during Summer Pre- and Post-Narmada Water Supplies							
	Bhuj Rural		Jamnagar Rural		Bhuj Urban		Jamnagar Urban	
Type of Domestic Water Use	Before	After	Before	After	Before	After	Before	After
Drinking and cooking	2,445	2,453	2,471	2,536	2,254	2,405	2,543	2,565
Washing	6,493	6,528	9,250	9,250	7,026	7,088	8,783	8,781
Bathing	9,865	9,592	18,069	17,997	12,489	12,548	18,227	18,080
Toilet use	6,424	6,424	9,435	9,435	8,339	8,260	9,997	11,603
Cleaning utensils	3,437	3,468	5,937	5,937	4,593	4,593	5,526	5,526
Scooter/car washing	0.00	0.00	0.00	0.00	383	383	413	413
Gardening	0.00	0.00	0.00	0.00	750	750	–	–
Floor cleaning	2,712	2,712	3,727	3,727	3,433	3,433	3,667	3,667

Source: Authors' own analysis based on data collected through a primary survey in four locations.

Impact of Drinking Water Quality

Health outcomes of drinking water supply provision are more a function of the quality of water supplied (Esrey *et al.*, 1985, 1997; WHO/UNICEF, 2000). Analysis of the quality of water supplied for drinking and domestic uses pre- and post-Narmada drinking water supply scheme for different subsectors of household water use sector, viz., drinking and cooking, other domestic uses and livestock drinking in the three seasons shows some definite trends across locations. According to a large proportion of the households surveyed, the quality of water supplied for domestic uses prior to Narmada was of high TDS, the water from a variety of sources, but mainly coming public bore wells. Only a few households opined that the water was *soft* and free from salts (Table 7.10). The situation had changed completely after the introduction of Narmada canal-based piped water supply, with a very small fraction of the households (less than 5 percent) expressing concern over the physical quality of water, particularly high TDS. Most of the households expressed the opinion that water is soft. Several qualities are attributed to soft water, some among them being, the less amount of time taken to cook food, the quality of tea being prepared from the water, etc.

Table 7.10:
Water Quality Impacts of Narmada Canal-based Water Supply on Different Household Uses

Season	Type of Household Water Use	Water Quality	Rural (% to Total Surveyed HHs)		Urban (% to Total Surveyed HHs)	
			Bhuj	*Jamnagar*	*Bhuj*	*Jamnagar*
Before Narmada canal-based piped water supply						
Monsoon	Drinking and cooking	1. High TDS	56	53	60	62
		2. High Fluoride	–	–	–	–
		3. Soft water	4	6	1	
	Other domestic uses	1. High TDS	57	52	60	62
		2. High Fluoride	–	1	–	–
		3. Soft water	3	6	1	
	Livestock	1. High TDS	56	54	60	–
		2. Fluoride	–	–	–	–
		3. Soft water	4	5	1	–

(Table 7.10 Contd)

(Table 7.10 Contd)

Season	Type of Household Water Use	Water Quality	Rural (% to Total Surveyed HHs)		Urban (% to Total Surveyed HHs)	
			Bhuj	*Jamnagar*	*Bhuj*	*Jamnagar*
Winter	Drinking and cooking	1. High TDS	56	51	60	62
		2. High Fluoride	–	1	–	–
		3. Soft water	4	6	1	–
	Other domestic uses	1. High TDS	56	53	60	62
		2. High Fluoride	–	–	–	–
		3. Soft water	4	6	1	–
	Livestock	1. High TDS	56	1	60	–
		2. High Fluoride	–	–	–	–
		3. Soft water	4	5	1	–
Summer	Drinking and cooking	1. High TDS	56	51	60	62
		2. High Fluoride	–	1	–	–
		3. Soft water	4	6	1	–
	Other domestic uses	1. High TDS	56	51	60	62
		2. High Fluoride	–	1	–	–
		3. Soft water	4	6	1	–
	Livestock	1. High TDS	–	1	60	–
		2. High Fluoride	–	–	–	–
		3. Soft water	4	5	1	–
After Narmada canal-based piped water supply						
Monsoon	Drinking and cooking	1. High TDS	–	–	–	–
		2. High Fluoride	–	–	–	–
		3. Soft water	60	59	61	62
	Other domestic uses	1. High TDS	–	–	–	–
		2. High Fluoride	–	–	–	–
		3. Soft water	58	59	61	62
	Livestock	1. High TDS	3	6	–	–
		2. High Fluoride	–	–	–	–
		3. Soft water	57	53	60	62
Winter	Drinking and cooking	1. High TDS	–	–	–	–
		2. High Fluoride	–	–	–	–
		3. Soft Water	60	59	61	62

(Table 7.10 Contd)

(Table 7.10 Contd)

Season	Type of Household Water Use	Water Quality	Rural (% to Total Surveyed HHs)		Urban (% to Total Surveyed HHs)	
			Bhuj	Jamnagar	Bhuj	Jamnagar
	Other domestic uses	1. High TDS	2	–	–	–
		2. High Fluoride	–	–	–	–
		3. Soft water	58	59	61	–
	Livestock	1. High TDS	3	6	1	–
		2. High Fluoride	–	–	–	–
		3. Soft water	57	53	60	62
Summer	Drinking and cooking	1. High TDS	–	–	–	–
		2. High Fluoride	–	–	–	–
		3. Soft water	60	59	61	62
	Other domestic uses	1. High TDS	2	–	–	–
		2. High Fluoride	–	–	–	–
		3. Soft water	58	59	61	62
	Livestock	1. High TDS	2	6	1	–
		2. High Fluoride	–	–	–	–
		3. Soft water	58	53	60	62

Source: Authors' own analysis based on data collected through a primary survey in four locations.

Note: As per the Indian standards on drinking water quality: the desirable limit for TDS is 500 mg per l and maximum permissible limit is 2,000 mg per l; the desirable limit for fluoride is 1 mg per l and maximum permissible limit is 1.5 mg per l; and the desirable limit for total hardness, as $CaCO_3$, is 300 mg per l and maximum permissible limit is 600 mg per l.

Changes in Health Expenditure on Water-related Diseases

A direct social impact of improved water supply in terms of better quality water with higher dependability, one would expect in areas such as Bhuj and Jamnagar is on public health, as the health consequences of poor water supply could be quite serious. For instance, consumption of high TDS water could result occurrence of kidney stones in humans. Similarly, consumption of water containing high levels of fluorides could cause several water-related diseases, such as dental fluorosis, skeletal fluorosis and "hunchback," due to crippling effect, depending on the level of the concentration of fluorides in drinking water, though the ultimate effect on human body would be determined by the nutritional value of the food intake, the climate of the region, and age of the person consuming the water. Contaminated water can be a cause of jaundice, dysentery, etc. One simple indicator used here to analyze the public health impact is the health expenditure of families using drinking water supplies from the Narmada canal-based system during a year and how this has changed from pre Narmada to post Narmada.

Table 7.11 shows that there has been a striking reduction in the incidence of water-related diseases, which include kidney stone and dysentery, in both rural and urban areas of Bhuj and urban areas of Jamnagar, reflected in the reduction in the number of families and persons affected, the number of days of employment lost, and the total annual expenditure on healthcare. For instance, in the case of rural Bhuj, the loss of employment in terms of the number of days came down from 35 to nil. In the case of rural Jamnagar, it came down from 20 to 10. A significant amount of the expenditure on healthcare results from reduced operations for kidney stone. For instance, in the case of Bhuj town, the expenditure came down from ₹44,200 to ₹25,750 per annum. In the case of rural Bhuj, it came down from ₹5,200 per annum to nil. No case of jaundice was reported by the families surveyed for both pre and post Narmada situation and, hence, is not included in Table 7.11.

Table 7.11:
Changes in Health Expenditure Due to Water-related Diseases

Name of the Waterborne Disease	Details	Rural		Urban	
		Bhuj	Jamnagar	Bhuj	Jamnagar
Before Narmada					
	Sample size	60	59	61	62
Dysentery	No. of families affected	1		2	
	No. of members affected	1		2	
	Loss of employment (no.)	5		2	
	Expenses incurred (₹)	300.0		1,200.0	
Chikungunya	No. of families affected			1	2
	No. of members affected			4	3

(Table 7.11 Contd)

(Table 7.11 Contd)

Name of the Waterborne Disease	Details	Rural		Urban	
		Bhuj	Jamnagar	Bhuj	Jamnagar
	Loss of employment (no.)			12	25
	Expenses incurred (₹)			7,000.0	6,000.0
Kidney stone	No. of families affected	1		4	
	No. of members affected	1		4	
	Loss of employment (no.)	30		36	
	Expenses incurred (₹)	5,000.0		36,000.0	
Others	No. of families affected		2		1
	No. of members affected		4		1
	Loss of employment (no.)		20		3
	Expenses incurred (₹)		900.0		100.0
After Narmada					
Dysentery	No. of families affected				
	No. of members affected				
	Loss of employment (no.)				
	Expenses incurred (₹)				
Chikungunya	No. of families affected			1	2
	No. of members affected			1	2
	Loss of employment (no.)			35	10
	Expenses incurred (₹)			750.0	6,500.0
Kidney stone	No. of families affected			1	
	No. of members affected			1	
	Loss of employment (no.)			8	
	Expenses incurred (₹)			20,000.0	
Others	No. of families affected	1		1	1
	No. of members affected	2		1	1
	Loss of employment (no.)	10		15	3
	Expenses incurred (₹)		1,500.0	5,000.0	100.0

Source: Authors' own analysis based on primary data collected from rural areas of Bhuj and Jamnagar, and Bhuj and Jamnagar towns.

Findings

- There has been a marked change in people's reliance on the water supply sources after the introduction of Narmada canal-based drinking water supply. With the introduction of Narmada water, communities in rural areas of Jamnagar started using only Narmada water source for drinking, cooking, domestic uses, and livestock drinking, whereas they used to depend on public bore wells and common stand posts prior to the introduction of Narmada waters. A similar trend was also observed in Jamnagar town, rural Bhuj, and Bhuj town. No significant improvement in the frequency of water supply in post Narmada period is noticed, in spite of major improvement in water availability situation in terms of dependability. In both rural and urban areas of Bhuj, the frequency of water supply had actually reduced after the introduction of Narmada canal-based water supply.

• Physical access to water sources had improved after the introduction of Narmada waters. The survey showed that a larger number of households get water within the dwelling or within the premises after the introduction of the Narmada piped water supply scheme. In rural Bhuj, 53 percent of the sample households got water within their dwelling premises (as compared to only 33 percent in the pre-Narmada scheme). Similar trend was also found in the case of rural Jamnagar and Bhuj, and Jamnagar towns. After the introduction of Narmada water, the households save time on water collection. Time saving was slightly higher during summer season, particularly in the case of rural Bhuj and Jamnagar town, where the magnitude and extent of water shortage used to increase during summer prior to the Narmada piped water. Overall, the reduction in time saving was only marginal because of the larger number of trips that family members had to undertake to collect water as a result of reduction in the frequency of water supply. The communities were able to productively use the time saved from water collection. In rural Bhuj and Bhuj town, 60 percent of the women who could save time in water collection were able to use the time for reaching the worksite on time. Another 40 percent reported to have used the time in earning additional income.

• While there is no significant water use at the household level for domestic and livestock purposes in volumetric terms, in post-Narmada times there was a marked improvement in the quality of water supplied. The quality improvement is crucial for places such as Bhuj and Jamnagar, which used to face serious problems of water quality, with public health consequences. But what is equally important is the fact that adequate quantity of water is being supplied from the single source of Narmada scheme—50.2 lpcd in rural Bhuj, 73 lpcd in rural Jamnagar, 60.7 lpcd in Bhuj town, and 70.3 lpcd in Jamnagar town—to meet all domestic requirements. As research at the global level has shown, such quantities of household water use are capable of bringing about positive health benefits, if water is of good quality. The quality of water supplied by Narmada-based piped water supply was perceived to be good in terms of physical and chemical features. Most of the sample households expressed the opinion that water is soft and low in TDS. Water quality improvement resulted in some perceptible reduction in the average annual expenditure on health, lower incidence of water-related health problems, and reduction in the number of days of employment loss due to water-related sickness. For instance, in the case of rural Bhuj, the loss of employment in terms of the number of days for the sample households came down from 35 to nil. In the case of rural Jamnagar, it came down from 20 to 10. In the case of rural Bhuj, the health expenditure came down from ₹5,200 per annum to nil.

***

Many water-scarce regions in India are facing acute shortage of water for domestic purpose during summer months. Problems are rampant in urban (Kumar, 2004; Mukherjee *et al.*, 2010) as well as rural areas (Kumar, 2004). Saurashtra and Kutch fall in this category. This is in spite of the fact that domestic water use accounts for a small fraction of water use, a large share goes to irrigating crops. This phenomenon further accentuated by the excessive dependence on groundwater, which is in the open access regime, for domestic water supplies through open wells, hand pumps, and bore wells. Since the farmers excessively depend on groundwater for meeting their agricultural demands, domestic water supply has to compete with irrigation. In most of the naturally water-scarce regions, the total water demand from all sectors of use put together exceeds the renewable water resources available. In such regions, the limited resources are largely appropriated by irrigators to meet their demands during kharif and winter, leaving the aquifers almost dry during summer. Water-intensive agriculture is practiced at the cost of basic survival needs of communities.

In urban centers falling in such regions, the local bodies or the water utilities have moved away from local rivers and streams to well water to local reservoirs to regional surface water imports as both the total population and population density of urban centers increased. Recent research shows that the dependence on surface reservoir is very high for cities, with large cities getting more than 90 percent of their supplies from water import from surface reservoirs. Today, large reservoirs play a crucial role in meeting the fast-growing urban water demands in India (Mukherjee *et al.*, 2010).

However, there is little appreciation among rural water supply agencies of many provincial governments in India of the fact that groundwater-based sources cannot ensure sustainable water supply. As a result, state governments continue to invest in local village-based water supply schemes, under the premise that they being small could easily be run by local panchayats. The accent on "decentralization of rural water supply management" and the demand-driven approach under the sector reforms in rural water supply had also led to this situation. The not so encouraging experience with regional water supply schemes with respect to operation and maintenance had also prompted some state governments to give up such schemes. Here, the goal of "decentralized management" and community preference overrides sustainability considerations. Gujarat is the first state to recognize this fact about sustainability problems in groundwater-based rural water supply schemes and indulge in large-scale regional water transfers to supply water to regions which are experiencing perennial water problems.

Further research would require long-term surveillance, to see how the vulnerability of communities to problems associated with the lack of adequate quantity of water of sufficient quality changes over time. Some of the probable long-term impacts are improved nutrition and health status of children, better education of school-going children as a result of reduced dropout rates, improved management of the village environment, and improved livelihoods of rural households. It is expected that with access to dependable source of good quality water, the rural households would take up animal husbandry, kitchen garden, and backyard cultivation.

8

Indirect Impacts of Irrigation and Drinking Water Supply

We have seen in Chapter 2 that gravity irrigation through canals produce several positive externalities on groundwater regime in the locality. They can be affected through changes in water availability in wells in the command area for irrigation, changes in water table conditions and yield of wells, and changes in the availability and quality of groundwater for drinking and domestic uses. While increased groundwater recharge can increase the yield of agro wells and increase the area irrigated by individual wells, rising water table can reduce the energy consumption for pumping groundwater and reduce the incidence of well failures (Shah and Kumar, 2008). All these have positive implications for economics of irrigated agriculture.

The augmented recharge from canal seepage and return flows from irrigated fields can often reduce the concentration of minerals in native groundwater, which cause serious health hazards like fluoride and salinity. Continuous recharge from canals and irrigated fields can ensure good yield of drinking water sources, sustainability of which gets threatened by the depletion of aquifers owing to overpumping by irrigators. Improvement in the groundwater quality can affect drinking water situation positively in areas where poor quality groundwater is used for domestic water supplies. Improved well yields can improve the drinking water situation in areas where drinking water sources tap hard rock aquifers.

As a matter of fact, there is heavy dependence on groundwater for drinking water supplies in India, with 50 percent of urban water supply and 80 percent of the rural water supply requirements being met from groundwater. This includes areas where groundwater quality problems also exist (Kumar, 2007). Several developing countries of Africa and East Asia heavily depend on groundwater sources for rural water supplies through hand pumps and shallow wells. Improved groundwater availability and quality can reduce the public investment for ensuring fresh water supplies for domestic uses throughout the year. As a result, the indirect impacts can extend beyond agriculture sector.

While some of these externalities add to private benefits, some reduce public costs. For instance, the reduction in well failures and the improved yield of wells reduce the private costs of farmers, thereby raising the net private gain from irrigated crop production. But reduced energy consumption for groundwater pumping reduces the economic cost and not the private cost for the famers as they are not confronted with the marginal cost of electricity for pumping groundwater under the subsidy regime. Similarly, in the case of drinking water, improvement in the performance of water supply sources reduces the public cost of provision of water supplies. Besides this, irrigated agriculture produces its own positive externalities by raising the wage rates in the farm and nonfarm sectors, by increasing the demand for wage labor (Narayanamoorthy and Deshpande, 2003).

Now, in the regions which are facing severe domestic water shortage, assured water supplies can induce several positive externalities due to the impacts on the households which have been illustrated in Chapter 7. Some of these externalities are: increased opportunities for wage employment and increased leisure for women, both resulting from time saving. However, these externalities get affected depending on the overall socio-economic dynamic of the target households and the characteristics of the new water supply system. Augmented water supplies may not always save time, unless the households have greater access to the new sources when compared to the old ones. In addition, there is a large benefit of replacing groundwater-based sources by surface water source which is unique to this context. The use of pipelines for transferring surface water from one NMC to Saurashtra and Kutch involves much less use of energy as compared to pumping groundwater from the local aquifers. These indirect benefits are analyzed in this chapter.

Social Benefits from Narmada Irrigation

Changes in Groundwater Availability and Water Level Fluctuations in the Region

Results of the analysis of data on the SWLs, for observation wells spread across the SSP command, from SSNNL for the period from 1996 to 1998 (pre-Narmada) and 2004 to 2009 (post-Narmada) were discussed in Chapter 5. As these results showed, there was a marked improvement in the groundwater level in the command areas post-introduction of Narmada canal water as compared to the pre-Narmada period of 1996–1998. By and large, trend in the SWLs in all the districts except Kheda is not only upwards but also higher when compared to the pre-Narmada period. Unlike what is generally observed in surface irrigation commands, groundwater use is quite extensive in the SSP command, resulting from conjunctive use of surface water and groundwater. The rise in water levels has helped the farmers in the command area, as there is reduction not only in the energy requirement for pumping a unit volume of groundwater, but also in the cost of deepening of wells. Also, there was marked improvement in the yield of wells, resulting in increased irrigation potential of wells (Figure 8.1). Subsidized electricity in the farm sector also reducing the economic cost of electricity for pumping of a unit volume of groundwater reduced.

Positive Externalities of Narmada Irrigation

Energy Saving Benefits for Well Irrigators in the Command Area

Our analysis shows that on average the depth to the water level in the sample wells surveyed in the command area had reduced across seasons. The results showed that there has been significant and consistent rise in

Figure 8.1:

Fluctuation in the Water Level in Wells (ft), Pre- and Post-Narmada

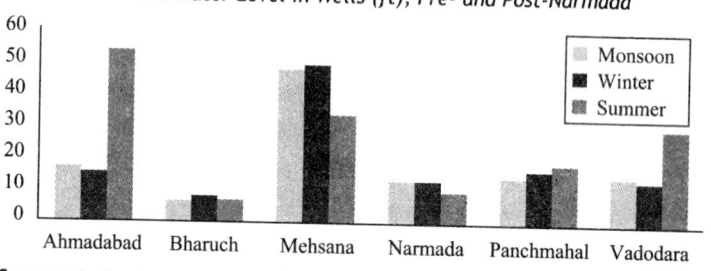

Source: Authors' own analysis of data based on the primary survey.

the water level across seasons after the introduction of Narmada waters, which first happened in 2002 in parts of the command in south Gujarat, particularly in Bharuch and Narmada districts. This change might have occurred due to the following reasons: (1) Introduction of canal water generally reduces the dependence on groundwater for irrigation, thereby affecting the reduction of the groundwater draft; and (2) groundwater recharge due to return flow from gravity irrigation using canal water. In addition to all these, the high rainfall received in most of these areas during the period from 2001 to 2006 might have influenced the water level trends with higher magnitudes of recharge. This being the case, on the other hand, groundwater use in backward, tribal areas of south Gujarat has actually increased over time with more and more farmers belonging to the scheduled caste and scheduled tribes investing in wells under government schemes, in areas which are not receiving canal water from Narmada. This might have had an influence on the regional groundwater level trends. In a nutshell, the groundwater situation in the canal command areas has apparently improved.

Every 1 m of reduction in pumping depth reduces the energy use for pumping 1 m³ of groundwater by nearly 0.055 kWh. Based on this conversion ratio, the reduction in pumping depths ($R_{P-DEPTH}$), the average volume of irrigation water applied per ha of well-irrigated area ($V_{GROUND-WATER}$), and the net well-irrigated area in Narmada canal command in ha ($A_{WELL-IRRIGATION}$), the total energy saving can be estimated using Equation (11) in Chapter 3.

The average volume of groundwater pumped per ha by the farmers and the average reduction in groundwater pumping depth in each region were estimated and provided in Table 8.1. Table 8.1 also shows the energy saving in kWh per m³ of groundwater pumped, the energy saving per ha of groundwater pumping and the total economic benefit due to energy saving per ha of well irrigation.

As Table 8.1 indicates, while the energy saving per unit volume of groundwater was highest in Mehsana, which has deep water table conditions, the total energy saving per ha of the well-irrigated area is also highest for Mehsana as the volume of groundwater used per ha of land is one of the highest for this area (2,542 m³). The energy saving per ha is second highest for Vadodara (1,170.7 m³), which has the highest volume of groundwater use per ha of the well-irrigated area (3,610 m³). Here, we should keep in mind the fact that the groundwater use is not only a function of the climatic conditions, with arid climates demanding more water, but also the cropping pattern selected by the farmers. While in areas like Ahmedabad and Mehsana, aridity would demand more water for crops, the cropping pattern is such that the overall water requirement is much less as compared to south Gujarat, where water-intensive crops such as wheat and paddy are grown.

The economic value of energy saving benefits in well irrigation owing to the rise in groundwater levels in the command area (ECOBEN$_{SAVE - ENERGY}$) can be obtained by multiplying the values in the

Table 8.1:
Average Groundwater Use for Crop Production in the Well Command Area (m³ per ha)

Name of Location	Volume of Groundwater Use per Ha after Narmada Water	Average Reduction in Pumping Depth (m)	Energy Saving in Kilowatt Hour per m³ of Groundwater Pumped (C × 0.055)	Total Saving in Energy for Groundwater Pumping by Farmers per ha (kWh)	Total Economic Benefit per ha
Ahmedabad	866.0	8.54	0.47	406.8	2,034.0
Bharuch	1,247.8	2.24	0.12	153.5	767.5
Mehsana	2,542.6	13.12	0.72	1,834.0	9,170.0
Narmada	2,665.1	3.76	0.21	551.4	2,757.0
Panchmahals	1,383.3	5.08	0.28	386.8	1,934.0
Vadodara	3,610.0	5.90	0.32	1,170.7	5,853.5

Source: Author.

last column of Table 8.1, by the area irrigated by wells in the SSP command area in each district.

Saving in the Cost of Well-deepening in the Command Area

This is also evident from the fact that the incidence of well failures has reduced significantly in four out of the six locations post-Narmada canal irrigation. The reduction in well failure was found to be very remarkable in Vadodara district, from 9.5 per year to almost nil. In Ahmedabad, and Narmada districts, the extent of reduction was lower. In Panchmahals, there has been an increase in the incidence of well failures from almost zero to 4.0. On the other hand, the command area of wells has increased post-Narmada canal water in four locations (Figure 8.2). While it remained more or less the same in one location, that is, Bharuch, it reduced in one location, that is, Panchmahals (Table 8.2). One reason for the negative trend in Panchmahals vis-à-vis the incidence of well failures and well command area could be the substantial increase in the number of wells tapping the hard rock aquifers of the district, which perhaps is disproportionately higher than the additional recharge available from the small canal irrigated area, resulting in well interference. It can also be seen from the table that wherever the incidence of well failures reduced, the command area of wells increased showing the natural tendency, except in Mehsana.

Table 8.2:
Changes in the Incidence of Well Failures and Well Command Area in Narmada Command after the Introduction of Canal Irrigation

Name of Location	No. of Incidence of Well Failure in the Village per Year		Average Command Area of a Well (ha)	
	Pre-Narmada	*Post-Narmada*	*Pre-Narmada*	*Post-Narmada*
Ahmedabad	3.00	1.2	36.6	37.8
Bharuch	4.75	5.2	6.1	6.05
Mehsana	0.00	1.2	14.7	15.1
Narmada	2.40	1.5	2.39	3.14
Panchmahals	0.00	4.00	5.10	3.90
Vadodara	9.50	0.00	4.40	4.50

Source: Authors' own analysis-based primary data collected from sample farmers in the six districts.

Figure 8.2:
Average Well Command Area (ha) in Narmada Canal Command before and after Narmada Waters

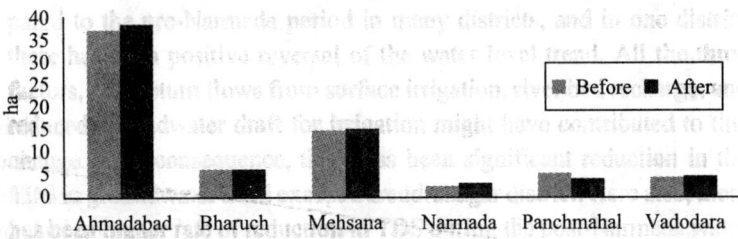

Source: Author.

Incremental Income of Well Irrigators in the Command Area

The incremental income from well irrigation in the command as a positive externality of canal irrigation can come from the following: (1) increase in area under well irrigation due to improved yield of wells; (2) shift in the cropping pattern toward water-intensive but high-valued crops, normally irrigated; (3) improved irrigation to the existing irrigated crops to obtain higher yields; and (4) reduction in the cost of irrigation due to the reduction in pumping depths and the reduced cost of well deepening. Another important source of income for the well-owning farmers is the increase in the extent of water sale.

First, we would examine whether there has been any change in the area of various crops grown by the well-owning farmers, who are likely to be benefited by canal irrigation. Table 8.3 shows the estimated average area under different crops in the command of sample well-owning farmers. The data covers all the three seasons, pre- and post-Narmada canal water. It shows that in most cases, the area has reduced, and in some cases, the area has increased. But, what is more notable is that the incidence of major reduction in area is more for rain-fed crops of kharif season, viz., green gram, paddy, chick pea, maize, and paddy, whereas the incidence of expansion in the area is more for irrigated crops—including those sown in kharif such as cotton and castor, and those sown in winter such as wheat in some locations—and water-intensive sugar-cane in Bharuch.

Synthesizing the data presented in Tables 8.2 and 8.3, one can see that the trend vis-à-vis cropped area is not in conformation with the trends

Table 8.3:
Change in Area under Various Crops (ha) in Well Commands after Narmada

Name of Crop	Change in Area under the Crop in the Well Command after the Introduction of Narmada Water in Canals in					
	Ahmedabad	Bharuch	Mehsana	Narmada	Panchmahals	Vadodara
Kharif						
Paddy	0.001		-0.09	-0.01	-0.47	-0.02
Jowar		-0.40	-0.08			0.01
Alfalfa	-0.005					
Cotton		-0.14	-0.04	0.09	0.47	-0.28
Green Gram		-0.03				
Chick Pea		-0.51				
Castor		-0.05	-0.06	-0.12	0.36	0.04
Sugarcane		0.00				0.12
Maize			-0.02	-0.02	-0.06	-0.02
Tomato			-0.01			
Bajra					0.01	-0.006
Groundnut				0.03		-0.006
Tobacco					-0.08	
Winter						
Wheat	-0.002	-0.07	-0.01		0.02	0.3
Alfalfa	-0.001		0.02		-0.02	-0.02
Tomato			-0.04			
Cumin			-0.03			
Fennel			-0.12			
Summer						
Maize				-0.12	-0.05	0.03
Bajra						0.04

Source: Authors own analysis based on primary data collected from sample well owning farmers in the Narmada canal command.

noticed vis-à-vis the command area and water level trends particularly in Ahmedabad and Mehsana. While the water levels have improved, and the well command area increased, the gross cropped area reduced slightly. This is perhaps because of the fact that farmers do not own well individually in such areas owing to very high capital cost of well construction. In these areas, over a period of time, the number of share-holders of tube wells has increased, with the result that the area irrigated by individual farmers reduced. Slight reduction in the average land holding size is also another reason for reduction in the size of the gross cropped area. Another possibility is that with greater reliability of groundwater resources, farmers replace the rain-fed crops by irrigated ones, but grow in smaller area. We would examine whether this has been the trend in the later part of this section.

There is significant increase in the area under irrigated crops in Panchmahals, with area under cotton and castor increasing, while area under paddy, which is mostly rain-fed, reduced (refer Table 8.4). Unlike Vadodara, Narmada, Ahmedabad, and Mehsana, where the underground formation is alluvial in nature, formations underlying the command areas in Panchmahals are consolidated rocks (GEC, 1997). In these areas, the groundwater potential is generally very poor, and therefore, any improvement in groundwater recharge through irrigation return flows from canals would bring about marked difference in groundwater table, making well irrigation more sustainable. Still, this benefit of improved recharge is not reflected in the command area, perhaps because the farmers are irrigating their crops more intensively or there is an overall increase in the number of wells in the area.

Nevertheless, one can say with great degree of confidence that this change in the cropping pattern is because of the improved availability of water in the wells, as the depth to the water level in wells has reduced in these areas. We would now examine whether there has been any change in the yield of crops. Table 8.5 presents the results of the analysis of change in the yield of crops grown in different seasons in the well command after the introduction of Narmada waters. Here, the positive values indicate increase in yield and the negative values, vice versa.

Table 8.5 shows that almost every crop grown in the well commands of the six locations showed yield changes, mostly positive after the intro-duction of Narmada waters. Only winter maize (in one location) and

Table 8.4:
Percentage Change in Area under Various Crops of Well Irrigators after Narmada Waters

	Ahmedabad	Bharuch	Mehsana	Narmada	Panchmahal	Vadodara
Paddy	1.47	–	–12.68	–20.0	–70.15	–5.88
Jowar	0	–100	–20.51	–	–	12.5
Kharif Alfalfa	–55.56	–	–	–	–	–
Cotton	–	–4.36	–5.88	25	671.43	–19.72
Green Gram	–	–30	0	–	–	–
Chickpea	–	–39.84	–	–63.16	0	13.33
Castor	–	–31.25	–8.33	–	150	240
Kharif Maize	–	–	–100	–9.52	–10.91	–66.67
Tomato	–	–	–10	–	–	–100
Kharif Bajra	–	–	0	–	9.09	–60
Groundnut	–	–	–	300	–	–60
Tobacco	–	–	–	–	–20.51	–100
Wheat	–2.44	–13.46	–1	0	3.33	1,000
Winter Alfalfa	–20	–	40	–	–40	–
Tomato	–	–	–40	–	–	–
Cumin	–	–	–23.08	–	–	–
Variyali	–	–	–46.15	–	–	–
Summer Maize	–	–	–	–92.31	–38.46	7.89
Summer Bajra	–	–	–	–	0	36.36

Source: Author.

green gram (in another location) showed some reduction in the yield. The only two crops which did not show any change in the yield were groundnut and summer maize. In the case of all other crops, they all recorded changes at least in one location, and in some cases more than four locations. The percentage increase in the yield was as high as 328 percent for castor in Bharuch. The increase in the yield of cotton was a lowest of 27 percent for Mehsana and highest of 60.9 percent for Narmada district.

Table 8.5:
Change in the Yield of Crops Grown in the Well Commands in Six Locations, Due to Narmada Waters

Name of Season	Name of Crop	Change in Crop Yield in Well-irrigated Command (kg per ha) in					
		Ahmedabad	Bharuch	Mehsana	Narmada	Panchmahals	Vadodara
Kharif	Paddy	629.42		401.9	1,533.0	1,626.9	1,387.3
	Cotton		404.0	683.9	750.6	526.6	735.2
	Chick Pea		426.6		179.3	56.9	375.0
	Jowar		895.5				
	Green Gram		1,226.4	−87.6			
	Castor		2,014.8	1,051.4		375.4	554.8
	Sugarcane						
	Maize			1,314.0	1,083.7	751.9	1,606.0
	Bajra			1,022.0		1,314.0	
	Tomato			17,695.2			
	Tobacco					178.12	
	Banana						28,470.0
Winter	Wheat	535.8	1,187.0	817.1	1,305.2	721.61	759.2
	Cumin			386.4			
	Fennel			512.5			
	Tomato			34,711.5			
	Maize				−2,504.1	1,314.0	1,668.1
Summer	Jowar	438					
	Bajra					949	1,200.9
	Maize						

Source: Authors' own analysis based on data collected through a primary survey.

In the case of paddy, the highest yield increase of 90.9 percent was found in the case of Panchmahals. It is quite likely that the factors which we have explained about the yield increase in the case of canal irrigators might have played a significant role in the case of well irrigators also, particularly the inputs. The empirical data collected from the field support this. For instance, our analysis shows that the intensity of groundwater use (m^3 per ha of cropped land) in the well commands had increased post-Narmada, clearly indicating that either farmers are applying more water to their irrigated winter crops or applying supplementary irrigation to the crops, which are otherwise grown under rain-fed conditions (Figure 8.3). This obviously could change the yields.

Therefore, in order to know how this has impacted on the overall net returns from these crops, it is important to examine how the cost of inputs has changed for these crops. Table 8.6 shows that the cost of cultivation has increased for all the crops except for tobacco in Narmada district, in which case the yield did not show any improvements.

Table 8.7 shows the changes in the net income from the crops. It shows that for all the crops, the farmers' net income from crop production has increased. The highest income increase was found in the case of cotton, with the values ranging from the lowest of ₹42,212 per ha in the case of Bharuch to the highest of ₹71,030 per ha for Vadodara district. This is followed by castor, in which case the highest increase in the net income was to the tune of ₹50,734 per ha in the case of Mehsana district, which is famous for oil seed production. In the case of green gram, a slight reduction in per ha income was found. It is important to remember that this crop generally grows under residual soil moisture and

Figure 8.3:
The Intensity of Groundwater Use (m^3 per ha) in Well Commands

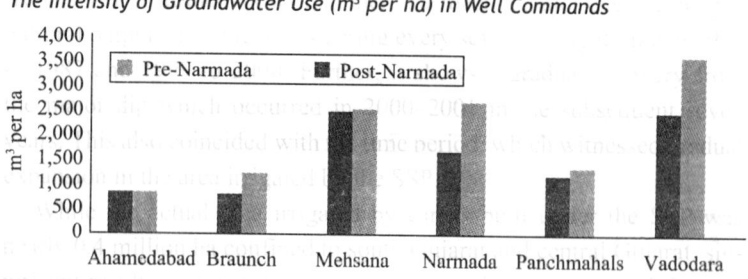

Source: Author.

Table 8.6:
Change in the Cost of Cultivation of Crops Grown in the Well Commands in Six Locations, Due to Narmada Waters

Name of Season	Name of Crop	Change in the Cost of Cultivation in Well-irrigated Command (₹ per ha) in					
		Ahmedabad	Bharuch	Mehsana	Narmada	Panchmahals	Vadodara
Kharif	Paddy	8,301.8		6,559.6	13,369.9	6,976.7	10,603.8
	Cotton		15,667.7	16,294.0	18,685.2	15,843.7	13,259.4
	Chick Pea		8,260.8		1,901.32	7,270.8	10,859.7
	Jowar		8,124.0				
	Green Gram		10,873.4	8,322.0		12,501.8	13,234.9
	Castor		16,381.2	17,131.7			
	Sugarcane				13,668.1	12,296.8	13,826.2
	Maize			5,256.0		10,455.1	
	Bajra			5,259.6			
	Tomato			24,579.1			
	Tobacco				−18,888.8		
	Banana					20,118.8	78,861.9
Winter	Wheat	8,580.2	10,501.1	6,995.2	12,114.6	7,844.8	9,374.7
	Cumin			11,742.8			
	Fennel			7,870.9			
	Tomato			6,099.1			
	Maize						
Summer	Jowar	2,518.5					
	Bajra						
	Maize				11,782.5	8,213.0	10,402.0

Source: Authors' own analysis based on data collected through a primary survey.

Table 8.7:
Change in the Net Income of Well-owning Farmers from Crop Production

Name of Season	Name of Crop	Change in Net Income from Crop Production in Well-irrigated Command (₹ per ha) in					
		Ahmedabad	Bharuch	Mehsana	Narmada	Panchmahals	Vadodara
Kharif	Paddy	-23,467.4		-1,181.9	925.3	12,351.6	12,675.2
	Cotton		42,212.0	67,733.9	48,289.0	65,293.9	71,029.9
	Chick Pea		17,871.3		14,152.8	11,103.3	16,539.4
	Jowar		7,449.4				
	Green Gram		15,187.7	-2,847.0			
	Castor		102,755.0	50,734.6			
	Sugarcane					35,521.8	35,553.0
	Maize				11,657.7		
	Bajra			21,900.0		5,310.75	12,599.5
	Tomato			14,085.3		10,087.14	
	Tobacco			187,851.0			
	Banana						
Winter	Wheat			17,285.5		29,448.2	41,587.6
	Cumin	4,408.3	6,478.1	49,316.4	5,006.8	5,851.0	4,506.0
	Fennel			54,708.4			
	Tomato			806,391.0			
	Maize						
Summer	Jowar			11,169.0	6,727.6		52,131.4
	Bajra	-13,906.5					
	Maize						

Source: Authors' own analysis based on data collected through a primary survey.

does not require much agronomic inputs, as a result of which, it is not the one which would show significant changes in yield and returns due to improved water availability in wells. In the case of summer bajra, a major reduction in the net income was found. But one can also see that this crop did not show any improvement in yield either, while its average inputs costs went up.

An astonishingly high value in net income change (positive) (₹806,391 per ha) was found in the case of tomato in Mehsana. But such values are unrealistic as the farmers grow such crops in very small areas, and a significant increase in price of such crops, whose market price is quite volatile, can raise returns from a hectare of the crop. Nevertheless, the increase in the net income is quite substantial for most crops, across locations. Table 8.8 shows the percentage change in the net income from various crops.

One might tend to argue that a part of the increase in income is because of the reduction in the cost of irrigation owing to the reduced cost of pumping groundwater, considering the change in the gross income and change in input costs for estimating the change in the net crop and farm income of farmers (this section), and then, separately estimating the cost saving in electricity use for groundwater pumping as another social benefit would result in double counting. But it is important to note here that the farmers are not paying for electricity on pro rata basis and as a result do not get the benefit of energy saving. Because of this reason, the energy saving benefit does not get translated into cost saving for them, and hence nullifying the chances of double counting.

In addition to the gain in the net income from existing crops, there are additional gains due to some new cash crops which were introduced by farmers in their well commands, after the introduction of Narmada canal. An example is sugarcane, which the well owners of Bharuch introduced newly in their commands. The crops is estimated to give a the net income return of ₹48,290 per ha of crop. This boosts the overall income from crop production further.

Change in Dairy Production in Well Commands

Improved recharge of groundwater, without a proportional increase in the number of wells, resulting in an effective increase in the command

Table 8.8:
Percentage Change in the Net Income of Well Irrigators from Crops after Narmada

	Ahmedabad	Bharuch	Mehsana	Narmada	Panchmahal	Vadodara
Paddy	114.96	–	-4.76	19.86	117.5	107.47
Cotton		264.14	206	587.1	397.15	250.91
Chickpea		489.18	–	294.93	222.37	524.47
Kharif Jowar		83.9				–
Green Gram		500.72	-48.15			–
Castor		2346	222.39	-100	307.7	491.95
Sugarcane						–
Kharif Maize			212.77	260.12	59.51	123.73
Kharif Bajra			135.64		385.76	-100
Kharif Tomato			87.78			-100
Kharif Groundnut						-100
Tobacco						–
Banana					144.69	74.3
Wheat	83.18	239.87	202.52	196.44	118.57	61.82
Cumin			324.9			–
Variyali			134.49			–
Winter Tomato			1,909.33			–
Winter Maize			–	123.01	443.48	377.04
Summer Jowar	-44.72					–
Summer Groundnut				-100		–
Summer Bajra					115.11	115.74
Summer Maize				–		-100

Source: Authors' own estimation based on the data collected primary survey.

area of wells—as found in the case of Vadodara, Bharuch, Narmada, Ahmedabad, and Mehsana—can potentially influence the animal husbandry activities of the well owners, the reason being the increased availability of biomass from the farm, in the form of dry and green fodder, and increased availability of water from the wells for livestock drinking. Our analysis shows (Table 8.9) that the holding of indigenous cows (in milk) increased in Ahmedabad, Mehsana, and Narmada, and marginally in Panchmahals and Vadodara, whereas it reduced in Bharuch. As regards crossbred cows, the holding size increased significantly in Vadodara, and marginally from nil in Bharuch. A reduction (from 0.11 to 0.05) was seen in Ahmedabad. As regards buffalo, its population size increased in Ahmedabad, Mehsana, Vadodara, and Narmada, while it reduced in Bharuch and Panchmahals. Overall, the trend in terms of livestock holding appears to be positive in most locations, except Bharuch. This coincides with the general trend in the cropped area, wherein the gross cropped area under the well command increased everywhere except Bharuch.

The absence of any change in livestock composition of farmers in Bharuch needs some explanation. In the case of Bharuch, the biggest change in cropping system is in terms of shift from short-duration rainfed varieties of cotton and castor to long-duration irrigated varieties and supplementary irrigation of crops, viz., jowar, chick pea, and green gram. Along with these, there has been reduction in the area of all these crops. Though the yield of these crops has improved, the overall output may not have increased. That said, the positive externality of canal water on well irrigation cannot be assessed in relation to what happens for an individual well command as the number of wells has increased over the years.

The average income of dairy farmers has also changed significantly after the introduction of Narmada canal water (Table 8.10) for all livestock types in all locations, except for indigenous cows in Bharuch and Mehsana. The positive trend vis-à-vis buffaloes and crossbred cows can be attributed to the increased production of crops which yield dry or green leafy biomass, such as paddy, jowar, wheat, bajra, and green gram, which are used as fodder,[1] and sometimes those which have by-products that can be used as feed for animals such as green gram and groundnut,

[1] While paddy, wheat and bajra provide dry fodder, jowar provides green fodder.

Table 8.9:
Average Size of Animal Holding of Well Owners before and after Narmada Canal

Name of Location	Indigenous Cow				Crossbred Cow				Buffalo				Bullock		Sheep	
	In Milk		Calf		In Milk		Calf		In Milk		Calf					
	Before	After	Before	After	Before	After	Before	After	Before	After	Before	After	Before	After	Before	After
Ahmedabad	0.11	0.20	0.07	0.08	0.11	0.05	0.10	0.08	0.84	0.89	0.18	0.20	0.00	0.00	0.00	0.00
Bharuch	0.10	0.05			0.00	0.10			0.19	0.14	0.10	0.10			0.19	0.00
Mehsana	0.01	0.08	0.00	0.00					0.16	0.55	0.05	0.21	0.01	0.02		
Narmada	0.03	0.05	0.00	0.05					0.43	0.57	0.19	0.26	0.72	0.84		
Panchmahals	0.00	0.10	0.10	0.10					0.52	0.45	0.21	0.24	0.93	1.02		
Vadodara	0.03	0.04	0.00	0.04	0.05	0.11	0.14	0.21	0.24	0.75	0.10	0.42	0.30	0.78		

Source: Authors' own analysis of the data collected through primary survey.

Table 8.10:

Average Income of Well-owning Farmers from Dairy Production after Narmada Canal (₹ per Animal per Lactation) before and after Narmada

Type of Animal	Name of the Location					
	Ahmedabad	Bharuch	Mehsana	Narmada	Panchmahals	Vadodara
Before Narmada						
Indigenous cow	5,178.74	5,420.00	13,187.50	4,680.00	0.00	2,002.50
Crossbred cow	8,255.38	0.00	0.00	0.00	0.00	11,740.00
Buffalo	13,473.57	2,000.00	15,017.60	8,750.37	5,018.38	5,941.14
After Narmada						
Indigenous cow	15,859.38	5,215.00	12,805.90	11,950.00	13,080.00	7,438.13
Crossbred cow	14,567.34	0.00	0.00	0.00	0.00	18,340.00
Buffalo	24,921.93	12,475.00	24,571.40	13,722.50	12,033.70	15,255.00
Change in Income						
Indigenous cow	10,680.64	−205	−381.6	7,270.0	13,080.0	5,435.63
Crossbred cow	6,311.96	0	0	0	0	6,600.0
Buffalo	11,448.36	10,475.0	9,553.8	4,972.13	7,015.32	9,313.86

Source: Analysis based on primary data.

as seen in the earlier section dealing with the impact on crop yields of well irrigators. Another important factor that might have contributed to the increased yield of dairy animals is the better care they might be getting with a general improvement in the welfare of the farmers keeping these animals. But the reduction in the net income in the case of indigenous cows may be because of lesser interest farmers probably show in rearing low-yielding indigenous varieties, which are more hardy and preferred in situations of water and biomass shortages.

Positive Externality of Canal Irrigation on Well Irrigators

The estimates of the positive externality induced by canal irrigation on well irrigation were made using Equation (11) presented in Chapter 3. The values of different variables that are required to estimate these indirect impacts per ha of the gross well-irrigated area ($WELL_{EXTERN-UNIT}$) is provided in Table 8.11. These estimates do not include Surendranagar, as the region's groundwater is totally saline and with that irrigation from Narmada is not expected to make any positive impact on the region's saline aquifers in the medium term.

It can be seen from Table 8.11 that while the average area under well irrigation had declined in four out of the six cases, the average net income from crop production substantially increased, with the average incremental income often exceeding the average net income prior to Narmada waters. Here, it must be mentioned that the average net income is the weighted average of the net income from various crops, estimated on the basis of the area under each one of the crops. Therefore, the shift in the cropping pattern toward high-valued, irrigated crops has definitely influenced the overall net income from unit area under crops. On the other hand, dairying can also influence farm surplus. There has been a marginal increase in the number of livestock, and a substantial increase in the average net income from dairy production per unit of livestock kept by the well irrigators, post-Narmada water introduction.

It is to be kept in mind here that it is theoretically incorrect to consider that improved groundwater balance would always translate into increased irrigation potential of groundwater and actual well-irrigated area in the canal command, when the positive groundwater balance is merely because of reduced draft and not because of increased recharge due to

Table 8.11:
Changes in Farm Surplus of Well Irrigators in Canal Command Areas

	Ahmedabad	Bharuch	Mehsana	Narmada	Panchmahals	Vadodara
Sample size	61	21	63	576	29	60
Current average gross irrigated area (ha per farmer)	0.17	4.47	3.80	0.87	3.50	3.05
Pre-Narmada avg. gross irrigated area (ha per farmer)	0.17	5.67	4.28	1.01	3.32	2.95
Avg. increase in gross irrigated area (ha per farmer)	−0.01	−1.20	−0.48	−0.14	0.18	0.10
Current average net irrigated area (NIA), well (ha per farmer)						
Kharif	0.08	4.02	2.41	0.79	2.34	2.14
Winter	0.08	0.45	1.36	0.08	0.73	0.76
Summer	0.00	0.00	0.03	0.00	0.43	0.15
Current gross well-irrigated area in the region: in SSP CA						
Avg. net income from well-irrigated crop (pre-Narmada): $\varnothing_{WELL}$	11,320.3	10,923.2	24,151.4	5,852.2	9,451.8	17,966.4
Average increase in net income from well-irrigated crops (₹ per ha): $\varnothing'_{WELL}$	9,814.1	34,836.7	45,249.5	28,703	18,212.3	36,228
Average livestock holding (present) (no. per ha of GWIA): N_{WELL}	8.82	0.09	0.23	2.03	0.55	0.77
Avg. net income per livestock (₹ per Unit): pre-Narmada: β_{WELL}	12,071.3	3,179.3	14,910	8,484.9	5,018.4	6,478
Increase in average net income per livestock (₹ per Unit): β'_{WELL}	11,088.4	7,664.5	8,292.2	5,157.4	8,118	8,809.8
Average increase in no. of livestock per Ha of GWIA: ΔN_{WELL}	0.55	−0.04	0.17	1.29	0.31	0.59
Positive externality of canal irrigation on well irrigation per Ha of gross well-irrigated area ($WELL_{EXTERN-UNIT}$)	113,587.1	32,466.9	46,640.7	49,176.3	24,719.0	47,422.6

Source: Authors' own estimates based on primary data analysis.

return flows. The reason is that the utilizable groundwater resources do not get enhanced there. Therefore, reduction in well-irrigated area found in our survey is quite natural in at least some of the locations as the improved groundwater balance which has occurred in those areas might be because of reduced pumping for irrigation, with tube well/purchased water irrigation being replaced by gravity irrigation or irrigation through canal lift.

Incremental Income of Wage Laborers Owing to Wage Increase

Empirical studies in the past have shown that increase in the irrigated area per unit agricultural labor force pushes the agricultural wage rates in rural areas by pushing the labor demand (Narayanamoorthy and Deshpande, 2003). We have earlier seen that canal irrigation increased the labor demand in the areas which are directly receiving irrigation water either through gravity or through canal lift, as reflected in the number of days of employment received by wage laborers in a year.

Increased demand for labor would ideally push the wage rates for casual labor both in farming sector and nonfarm sector, particularly in areas where there is already good demand for labor in these sectors (Narayanamoorthy and Deshpande, 2003). All the six districts we have surveyed fall under this category in view of their proximity to the cities and towns, and therefore, the demand for unskilled labor in the nonfarm sector is generally high. Further, agriculture is generally prosperous in these regions. The only exception is Panchmahals district, which has its agriculture still very backward owing to the peculiar sociocultural environment and a still low extent of irrigation from Narmada canals. Table 8.12 shows the current wage rates and increase in wage rate (after Narmada water introduction) in the six locations. Amongst the six locations, Bharuch recorded the highest growth in wage rates in the agriculture sector. It is even higher than that in the command areas of Ahmedabad which is quite close to that of the metro. The disproportionately high wage rates in areas such as Bharuch and Narmada can only be attributed to the intensive irrigated agriculture being practiced in these areas, due to the presence of Narmada canals.

Having stated this, it is important to remember that rural wage rates for casual labor have gone up all over the state of Gujarat in the past few

Table 8.12:
Changes in Wage Rate (₹ per Day) of Farm Laborers in the Five Locations over Time across Seasons for Male and Female Agricultural Laborers

Name of Location	Current Wage Rates for Men			Current Wage Rate for Women		
	Monsoon	*Winter*	*Summer*	*Monsoon*	*Winter*	*Summer*
Ahmedabad	97.8	96.7	86.7	87.9	95.6	86.7
Panchmahals	53.8	53.8	53.8	48	48	47.5
Bharuch	91.6	91.6	91.6	91	91	91
Mehsana	91.4	91	89.5	90.7	90.7	84
Narmada	81.6	80.6	67.7	72.6	77.2	72.5
Vadodara	53.4	55.6	58	49.4	59.5	48.4
Change in Water Rates after Narmada						
Ahmedabad	44.3	43.2	54.2	35.1	40.5	59.2
Panchmahals	21.87	21.9	21.9	17.5	17.5	22.5
Bharuch	56	56	56	54.7	54.7	54.7
Mehsana	38.5	38.1	33.8	36.3	36.3	24.5
Narmada	34.7	33.7	30.1	24.8	28.2	36.9
Vadodara	20.5	22.5	22.5	18.6	28.5	13.5

Source: Authors' own analysis based on data collected through a primary survey.

years (Hirway and Shah, 2011), because of the inflationary effect and also the effect of the National Rural Employment Guarantee Scheme, which offers 100 days of labor in every village in a year, though they have not kept pace with the national level trends. For instance, the wage rates for casual male labor went up from 43.9 to 68.5 per day during 1999–2000 to 2007–2008, a period of eight years in the state. Similarly, the wage rate for casual female workers went up from ₹34.4 per day to ₹58.9 per day during the same period. This shows an annual growth rate of 5.7 percent for male workers and 7 percent for female workers (Hirway and Shah, 2011).

Therefore, it is necessary to factor out these effects while estimating the real increase in wage rates in the farm sector owing to irrigation from Narmada. We assume here that the conditions are quite unfavorable for wage rates to go up in Panchmahals as a result of Narmada irrigation, for the reasons explained above, and therefore, would consider the increase in wage rates found in that area as the effect of inflation and rural employment guarantee scheme, in conformation with the overall

Figure 8.4:
Estimated Effective Increase in Wage Rates in Real Terms (₹)

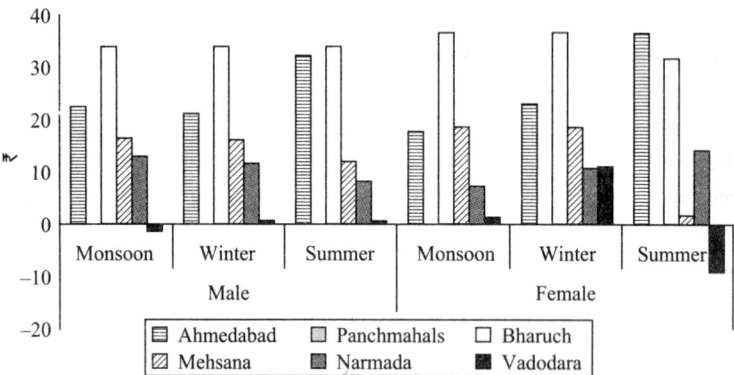

Source: Author.

trends in wage rate in Gujarat. Thus, the effect of irrigation expansion owing to Narmada was estimated by subtracting the wage rate increase observed in Panchmahals over the time period considered in our analysis from the incremental wage rates prevailing in the locations concerned. The estimates of the real increase in wage rates owing to Narmada irrigation in the five locations, in different seasons, for male and female laborers are presented in a graphical form in Figure 8.4.

An alternate method to understand the effect of irrigated area expansion on wage rate increase in Narmada command is to compare the wage rates in these areas against the average wage rates in Gujarat. The average wage rates in rural Gujarat for casual work (adjusted to a rate of inflation of 7 percent) in 2011 were around ₹83.3 per day for male workers and ₹72.3 per day for female workers,[2] whereas the average wage rate for casual workers worked out for the four district of Narmada, Bharuch, Ahmedabad, and Mehsana is ₹88.1 per day for male workers and ₹85.9 per day for female workers.[3] Hence, the wage rates for casual labor are much higher in the Narmada command area. These are conservative estimates, and the actual difference in wage rates between the Narmada

---

[2] Estimates by the authors based on Hirway and Shah (2011).

[3] Here, we have not considered the districts of Panchmahals and Vadodara, as the wage increase in those locations was disproportionately low.

command area and non-command area would be still larger. The reason is that the estimates of average wage rates for Gujarat mentioned above might have included some of the command area districts as well. The biggest gain in wage rate is for female workers with an average increase in wage rate of ₹13.3, whereas for male workers, the average gain is ₹4.8.

Further, large water infrastructure projects have multiplier effect on the economy of the region which they serve, through a variety of indirect benefits they produce. Though we have quantified many of these indirect benefits (well irrigation benefit, energy saving benefit, drinking water supply benefit, increase in wage rates in labor market, and clean energy benefit), the aggregate figures of some of these benefits could not be estimated in the absence of data on parameters such as the number of wells benefited by recharge in the canal irrigated area, number of wage laborers who got higher wages, etc., due to increased demand for wage labor in agriculture. Hence, we have used a "multiplier," equal to the lowest in the range (1.4–2.0) available from the published literature for large water projects, on average farm surplus from canal irrigation to obtain the total effect on the economy. The total effect (direct and indirect) of canal irrigation on region's economy was estimated to be ₹68.25 billion. This is based on the following: (1) an estimated farm surplus of ₹81,200 per ha of the gross irrigated area (gravity and lift irrigation), (2) a gross irrigated area of 0.60 m. ha by Narmada canal system, and (3) an assumed multiplier coefficient of 1.4. This means there was an indirect economic benefit (multiplier effect) of ₹32,500 per ha of irrigation, which produces a direct economic benefit of ₹81,200.

Reduction in the Cost of Domestic Water Supplies in the Command Area Villages and Towns

The pipeline network based on Sardar Sarovar Narmada canal and Mahi canal is expected to serve nearly 54 percent of the villages in Gujarat, apart from 125 towns.[4] The entire Saurashtra and Kutch and most parts of north and central Gujarat are to be covered under this scheme. But there are thousands of villages in south and east Gujarat which are not to be covered by this pipeline network. These areas are generally being

---

[4] Based on www.gwssb.org/pdf/narmadaprojects.pdf

considered as the well-endowed regions of the state in terms of water resources.

But even within south Gujarat, there are two distinct areas with respect to geo-hydrological environment. The first one falls in the alluvial belt covering parts of Vadodara, Narmada, Surat, and Bharuch. This area does not face problems of physical shortage of water during any part of the year, owing to the rich alluvial aquifers in the area which get sufficient replenishment from rainfall, and plenty of surface water in ponds, though they are problems of water quality. The other one falls in the hard rock region, with consolidated rocks of Deccan trap basalt. Parts of Vadodara, Narmada, and Bharuch, which fall inside the Narmada command, are in the Deccan trap area. Now, Panchmahals in east Gujarat is fully underlain by crystalline hard rock formations. In spite of moderate to high rainfalls, this hard rock region has poor groundwater potential (GEC, 1997), Drinking water shortage occurs in this region during the summer months, due to drying up of wells.

Like irrigation wells, the drinking water supply sources located inside the command area and which tap local aquifers are likely to be benefited by canal irrigation through return flows from irrigated fields and seepage from canals. These benefits get affected in the following way: (1) rise in water levels in the wells, which can reduce the cost of pumping groundwater, thereby reducing the operational cost of village panchayats running these schemes; (2) reduced incidence of well failures and, therefore, reduced investment for well deepening, etc.; (3) rejuvenation and improved yield of wells, resulting in increase in the level of supply of water, especially during summer months; and (4) dilution of the minerals present in the groundwater, thereby improving the quality of drinking water supplied by the local well-based schemes. While the benefits discussed under item "1" and "2" can be put in monetary terms, evaluating the third and fourth types of benefits would require estimation of water demand functions and this method is attempted here.

The economic benefit from the improved yield of drinking water wells and sustainable drinking water supply can be treated to be same as the amount of money required to transport water from a distant sources to meet the water supply needs during summer months. Further, the economic benefits from the improved chemical quality of water in drinking water wells can be treated to be the same as the cost of treating the poor quality water, through treatment processes like the desalination or defluoridation, depending on the case. While the first three types of impacts of canal water introduction would be more visible in hard rock regions, because of the unique nature of aquifers in those regions, the fourth type of impact, that is, dilution of minerals present in groundwater, is more likely to be observed in the alluvial areas of north Gujarat. But, unfortunately, all the villages surveyed for canal lift in Ahmedabad and Mehsana were found to have been covered by the Narmada canal-based pipeline scheme, and, therefore, the issue becomes nonexistent in these areas. Therefore, the social benefits of canal irrigation in the form of improved sustainability of groundwater-based drinking water schemes would be affected only in south and east Gujarat parts of Narmada command.

The reduction in pumping depth is estimated to be 12.5 feet, that is, 3.75 m (based on Figure 8.1). The cost of generating and supplying energy for groundwater pumping is taken as ₹5 per kWh. The volume of water required to meet the domestic water supply needs is estimated to be 36.5 m³ per capita for the whole year, and this is based on the assumption that the communities would require 100 liter of water per day. While it has to be met from wells during most parts of the year, during summer months, only part of this demand will be met from wells and water will have to be supplied though tankers at the rate of 40 lpcd in the absence of surface irrigation from Narmada canals. For tanker water, we have considered the ongoing market price, that is, ₹50 per m³. As a result, water delivery through Narmada canals would augment the water supplies in wells so as to avoid the dependence on tankers. Further, we have considered the population depending on groundwater-based drinking water supply sources, in the scarcity hit region, at present to be 5 million. Thus, the total economic benefit was estimated to be ₹1,069.4 million per annum (213.93 × 5 × 10⁶ = ₹1,093.3 million), as per Equation (12) in Chapter 3. The detailed calculations are presented in Table 8.13.

In addition to the benefits from the reduction in pumping depths and improvement in water supply quantity, benefits are also accrued from improvement in the quality of water, occurring due to dilution of groundwater. Research carried out in one of the villages in Vadodara district, which receives canal water for irrigation shows significant improvement

Table 8.13:
Indirect Benefit of Canal Irrigation through Improved Water Supplies

Type of Water Supply Source	Water Supply Requirement Met from the Source (lpcd)	Reduction in Pumping Head due to Canal Recharge (m)	Energy Saving per m³ of Pumping due to Reduction in Pumping Head by 1 m	Volume of Water Used (m³ per annum)	Cost Saving per m³ of Water	Economic Benefit per Capita
Bore well/ Open well	100 in monsoon and winter and 60 in summer	3.75	0.055	32.90	1.03	33.93
Tanker Supply	40 in summer	NA	NA	3.60	50.0	180.00
Total Cost Saving for a Population of five million people depending on Water Supplies						1,069.4 m

Source: Authors' own analysis based on primary and secondary data.

in the quality of groundwater used for domestic purpose. The location selected for the study falls in an area with underlying basalt formations. The community in the village is dependent on bore wells drilled on the banks of Orson River for domestic water supply. There has been a notable reduction in the number of families reporting problems of water quality in the sources and they attribute the quality improvement to dilution of minerals present in groundwater across the seasons. As Figure 8.5 shows, no household reported high TDS in the water used for drinking and cooking in their public water sources, which include the hand pumps in the locality after the introduction of Narmada canal water in the village, while at least five households had problems with their drinking water source, that is, the distant bore well located on the river bank) before Narmada water was introduced. The low incidence of poor water quality is because the bore well which is tapped for drinking water supply gets replenished continuously by the river, which carries water free from salts.

But the village households use water from local bore wells and hand pumps for domestic water supply and livestock drinking, which again is available through individual tap connections and stand posts. As Figure 8.6 shows, 25 out of the 30 families surveyed reported problems with their domestic water sources during winter and summer months

Figure 8.5:
Quality of Water Used for Drinking and Cooking

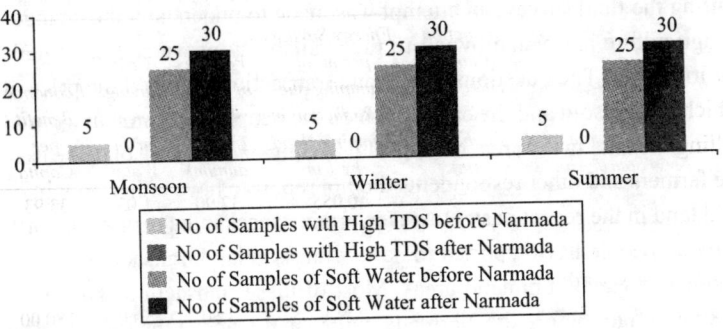

Source: Author.

Figure 8.6:
Quality of Water for Domestic Purpose

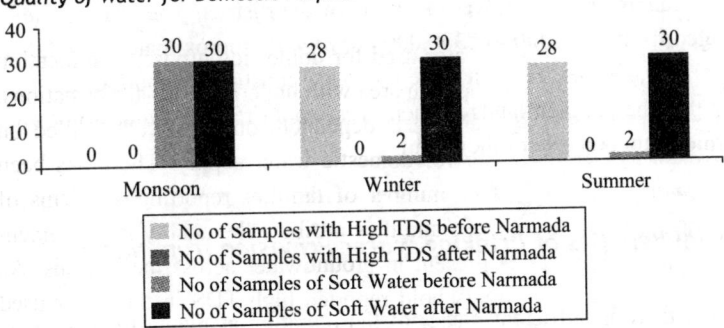

Source: Author.

prior to the introduction of Narmada canal water in the village, with no families complaining about the water quality during the rainy season. But no families reported the problems of salinity in water after Narmada canal water was introduced. This is quite possible because of the fact that the wells get recharged during monsoon with the freshwater remaining on the top of the water table. This water gets used up during the rainy season, with the water having salinity remaining in the lower strata of the consolidated formations and this water gets used up during winter and summer months. After the introduction of Narmada canal, the wells continuously get recharged by the return flows from canal irrigation.

Impact of Canal on Land Markets

During the field survey, an attempt was made to understand the changes in agriculture and value owing to the availability of water from canals for irrigation. The questions about land transaction included the price at which land is sold and the caste to which the people, who are buying and selling the land in the area, belong. But barring one or two isolated cases, the farmers and other respondents did not report to have either bought or sold land in the recent past. However, from Banaskantha district in north Gujarat, people have reported to have witnessed land transactions in the designated canal command areas. Most of these transactions happened 7–8 years ago, before the Narmada main canal reached the district. The farmers belonging to Patel community bought land in Wav and Tharad talukas from land owners belonging to Darbar (Thakur) and Koli communities, on hearing about SSNNL's plans to start the construction of main canals up to northwestern parts of the district. The price of land ranged from ₹100,000 to ₹300,000 per *bigha*[5] (approximately ₹400,000 to ₹1,200,000 per ha). But once the canal construction started, expecting that the parched lands in their area would receive canal water, the farmers stopped sale of their land.

Social Benefits of Drinking Water Provision from Narmada

While there is sufficient pieces of the literature available from around the world on the socioeconomic benefits of improved water supplies such as the reduction in time spent in water collection, reduced calorie required for fetching water, and saving in the cost of collecting water from alternative sources,[6] there is little understanding of the welfare effects, or the way improved water supply alters other economic production functions, though Whittington and Chao (1992) discussed several methods for estimating the economic value of improved drinking water

[5] Bigha is the unit of area in local language. 1 bigha ≅ 4 hectares.

[6] Based on Cairncross and Cliff, 1986, for Mueda in Mozambique; Chetwynd *et al.*, 1981, for Korea; Feachem *et al.*, 1977, for Lesotho; White *et al.*, 1972, for east Africa.

supply provisions such as estimation from conventional demand models, hedonic property value approach, and random utility model approach (see Whittington and Chao, 1992, pp. 40–55).

Changes in Wage Earning Potential of Women and Men Due to Increased Time Availability

As noted by Whittington and Chao (1992), it is often assumed that time savings in collecting water will have "economic value" only if men and women use them to undertake "economically productive work." This is, however, a misunderstanding of the economist's definition of "economic value." The time savings have economic value to the extent that households are willing to pay for them—whatever the reason be. In other words, if households are willing to pay to save time, what they do with the time saved should be of no concern insofar as the economist is interested in the economic benefits of the improved water system. Nevertheless, an attempt was made to see how women use the saved time. It was found that at least in the case of rural Bhuj and Bhuj town, 60 percent of women who could save time in water collection (30 or half of the total sample) were able to use the time for reaching the worksite on time. Another 40 percent also reported to have used the time in earning additional employment, while around 13 percent used the time for leisure (Table 8.14).

Table 8.14:
Positive Externality of Improved Water Supply

Name of Location	No. and % of Families Who Have Saved Significant Amount of Time in Water Collection	Number of Families Who Allocate the Saved Time for		
		Reaching Worksite on Time	Earning Additional Employment	Leisure
Rural Bhuj	30 (50)	18	13	4
Bhuj Town	15 (25)	10	2	0
Rural Jamnagar	4 (6.7)	0	4	1
Jamnagar Town	2 (3.3)	0	0	0

Source: Authors' own estimates based on a primary survey.

Reduction in the Use of Electricity for Groundwater Pumping for Water Supply

Gujarat is among the few states whose use of groundwater for urban and rural water supply is one of the lowest in the country. As far back as 2001, the level of groundwater use for municipal water supply in the state was only 7.5 percent. With the introduction of Narmada canal-based supply, this has further reduced. Narmada water supply is also extended to many thousands of villages in Gujarat and has replaced bore well-based decentralized water supply schemes. It is a well-known fact that groundwater resources in most parts of Gujarat face problems of poor chemical quality, manifested by high levels of fluoride, nitrates, and salinity (IRMA/UNICEF, 2001; Kumar, 2007). Therefore, overdependence on such sources of water supply poses serious public health risks (IRMA/UNICEF, 2001). Over and above, aquifers are showing increasing signs of overdevelopment in most of the naturally water-scarce regions of Gujarat, though the nature and degree of overexploitation vary from region to region. But overexploitation has serious consequences for the sustainability of well-based water supply schemes in rural and urban areas. During droughts, several hundreds of tube wells and bore wells are drilled in urban areas to tap water to meet the municipal requirements, but with very short lives, or water from neighboring irrigation reservoirs is being drawn, whereas in rural areas, as noted by Kumar and Talati (2000), the drought-proofing measures include sporadic drilling of wells and drawing water from irrigation reservoirs. These ad hoc measures do not, however, serve as lasting solutions for drinking water scarcity problems.

In the light of these, the state governments, over the years, had made systematic efforts to move away from local groundwater-based sources to improve the sustainability of water supply schemes and started building schemes that tap water imported from the Narmada basin. While many had questioned the logic of transporting water to such distant towns/cities and villages in different regions from Narmada canal in south Gujarat at prohibitive costs, two points need to be kept in mind: (1) Local groundwater resources do not provide freshwater that is free from TDS, or that is potable, throughout the year in most of the regions covered by the grid; and (2) the local groundwater resources, even if found to be of reasonably good quality, fail to meet the domestic water supply requirements of population during summer months, as aquifers get dried up by the end of the winter season. Hence, such a large investment for building infrastructure for water transport even to meet the peak summer demand is highly justifiable, as it support the lives of millions of people living in rural and urban areas.

The presence of large infrastructure, consisting of the water impounding reservoir, canals, and giant pipelines, for water transfer had enabled transport of water from water-rich Narmada basin to the water-scarce regions, raising the value of water significantly. While the opportunity cost of diverting water from Gujarat's allocation of 9.75 MAF in Narmada for domestic uses is much less than the opportunity cost of using water from local aquifers, the benefits are much higher than that accrued from the use of local groundwater owing to the water from Narmada being of better natural quality with no requirement of a major treatment. These double benefits are, however, not considered in our analysis.

But one significant positive externality induced by the surface water import is on electricity consumption in urban and rural water supply schemes, with the local draft of groundwater reducing drastically. As Table 8.15 indicates, the average cost of electricity for pumping and supplying surface water in urban local bodies (ULBs) of Gujarat stands at ₹0.89 per m³ of water supplied, against ₹2.64 per m³ of water in the case of groundwater-based schemes.

There are 9,633 villages in Gujarat, which are proposed to be covered by the water supply grid and using Narmada waters. Currently,

Table 8.15:
The Cost of Electricity for Water Supply

Indicator	Average Cost of Electricity for ULB Using Groundwater		Total Population to be Covered by Narmada (million)	Estimated Cost Saving in Energy Use for Water Pumping (million)
	For ULBs Using Groundwater	*For ULBs Using Surface Water*		
Unit Cost of Electricity	2.64	0.89	24.33	857.70

Source: Mehta and Mehta (2011) based on Performance Assessment System (PAS) Survey, 2009.

6,513 villages are getting water from Narmada canal through this grid. In addition, there are 125 towns and cities which are to receive water from Narmada canal-based water supply. All these cities/towns and villages are falling in semiarid and arid areas, with low to medium rainfall and deep groundwater table. It essentially means that these cities/towns and villages will have to depend on surface water resources if they have to meet their water supply needs—in terms of quantity and quality—on a sustainable basis.

The total volume of water required to be pumped from the aquifers, to meet the requirement during monsoon and winter season (V_{DRINK}) would be 1,343 million liter per day or 490 MCM per year. This is based on the estimates provided by Talati and Kumar (2005) for the service area of the entire grid covering a total population of 24.33 million people for the whole year.

The total economic benefit due to energy saving is, therefore, estimated to be ₹857.7 million per annum (Table 8.15) as per Equation (13) in Chapter 3. But the actual benefit would be larger in view of the fact that the actual population being planned to be covered by the scheme is larger.

Findings

- The introduction of surface water from Narmada canal for irrigation has resulted in changes in groundwater level trends over time, marked by relatively higher annual rise in water levels in the observation wells in the command area, except in Kheda district, during the post-Narmada period. Similarly, there has been a marked reduction in the salinity of groundwater over time, as indicated by relatively greater annual rate of decline in TDS during the post-Narmada period.
- Irrigation wells located in the command area of Narmada canal are found to have been benefited by seepage from canal water. This is manifested by consistent rise in groundwater levels across seasons, resulting in reduced depth of pumping. The highest average rise in the well water level during monsoon (47 ft) and winter (49 ft)

was found in wells located in Mehsana, and during summer (53 ft) in Ahmedabad. The rise in the water level could be explained by reduced pumping by non-well-owning farmers reducing, thereby, their dependence on water purchase and increased recharge from canal irrigation. Even well-owning farmers also appear to have reduced their dependence on wells for irrigating crops.

- There has been a notable reduction in the incidence of well failures in the canal command area (four out of six locations) and an increase in command areas of wells (four out of six locations) post-Narmada. The reduction in well failure was found to be very remarkable in Vadodara district, from 9.5 per year to almost nil today. But the same trend was not visible in Panchmahals district. A probable reason can be the substantial increase in the number of wells tapping the hard rock aquifers of the district over the past few years, which perhaps is disproportionately higher than the additional recharge available from the small canal irrigated area, resulting in well interference.

- After the introduction of Narmada waters, the cropping pattern of well irrigators in the canal command had changed; there was substantial increase in the yield and income returns per unit of land (₹ per ha) for well irrigators. While there were incidences of major reduction in the area for the "normally" rain-fed crops of kharif season, the area under irrigated crops such as cotton, castor, and wheat had increased in some locations, and the same result is visible for water-intensive sugarcane in Bharuch. Almost every crop grown in the well commands of the six locations showed yield changes, mostly positive after the introduction of Narmada waters. The percentage increase in the yield was as high as 328 percent for castor in Bharuch. For all the crops, the net income had increased substantially. The highest income increase was found in the case of cotton, with the values ranging from the lowest of ₹42,212 per ha in the case of Bharuch to the highest of ₹71,030 per ha for Vadodara district. The increase in the yield and income from crops in the well commands could be explained by the greater dosage of irrigation to these crops, which is reflected in the increase in the average volume of groundwater applied by the farmers per ha of land post-Narmada.

- There was a marginal increase in the livestock holding for the well irrigators post-Narmada water. Overall, the trend in terms of livestock holding appears to be positive in most location, except Bharuch.
- Increased demand for farm labor had resulted in substantial increase in wage rates in the selected locations post the Narmada water introduction, which was found to be disproportionately higher than the historical increase in wage rates in other areas of Gujarat that are not so much benefited by Narmada canals. Bharuch recorded the highest growth in wage rates in the agriculture sector, with an aggregate net increase of ₹34 in the case of male workers and ₹37 in the case of female workers. The exceptionally high wage rates in areas such as Bharuch and Narmada is attributed to intensive irrigated agriculture being practiced in these areas, due to the presence of gravity irrigation from canals.
- The indirect benefit from canal irrigation in the form of the reduced economic cost of energy used for pumping groundwater for irrigation was estimated to be huge. For every hectare of well-irrigated area in the SSP command, the economic benefit to the society through energy saving in groundwater pumping ranges from a lowest of ₹768 in Bharuch to ₹9,170 in Mehsana. Such huge variations occur because of two factors: (1) major variations in the rise in the groundwater table observed across regions, which cause proportional variation in the energy saving for pumping a unit volume of groundwater and (2) major variation in the amount of groundwater used per ha of well-irrigated area, owing to variations in climate. It is quite clear that Mehsana derives the maximum benefit because of the intensive groundwater use
- The indirect economic impact of canal irrigation in the form of improved sustainability of groundwater-based drinking water supply schemes in south and east Gujarat parts of Narmada command for a population of five million is estimated to be ₹1,032.5 million per annum. This is based on the assumption that the drinking water supply sources located inside the command area and which tap local aquifers are likely to be benefited by canal irrigation through return flows from irrigated fields and seepage from canals in the form of: (1) the reduced depth of water

table, which reduces the operational cost of the scheme and (2) the increased availability of water in the wells for drinking water supply during summer months, which saves the money spent for tanker water supply. The primary survey conducted in Vadodara district showed improvement in the quality of groundwater used for domestic water supply.

- The indirect economic impact of Narmada canal-based piped water supply to a total population of 24.33 million in Saurashtra, Kutch, and north and central Gujarat, owing to the saving in energy used for pumping groundwater for rural and urban water supply, is estimated to be ₹857.7 million per annum. This is based on the fact that with the introduction of Narmada canal-based supply, dependence on groundwater for urban and rural water supply has reduced substantially.

- The other positive externalities of the provision of reliable water supply of good quality was that the women members of the households, who are engaged in water collection, could use the time saved in collecting water for leisure and sometimes finding new employment.

- The total of direct and indirect economic benefits from canal irrigation was estimated to be ₹68.25 billion, considering a total net farm surplus of ₹85,200 from every hectare of irrigation, an economic multiplier of "1.4" for large surface irrigation projects on the region economy and a gross irrigated area of 0.60 million ha.

***

In the past, many large and medium water resources projects in the country had been critiqued for their poor performance in terms of the area irrigated. But too little attention has ever been paid to the other sectors of the economy which they alter. Our analysis of the SSP shows that it can change well irrigation economy, by increasing the net returns of groundwater irrigators in the command area, by saving the cost of electricity supplied for groundwater pumping, by saving the public expenditure on the provision of freshwater supplies, including that from alternate sources during lean season. Mainly, the enhanced economic output from sustainable well irrigation resulting from augmented recharge from canal seepage and irrigation return flows is too large to be ignored. The study reinforced the argument by Shah and Kumar (2008) that the quantitative criteria for evaluating the benefits of large water resource projects need to be broadened to capture the indirect benefits.

The study also shows that there is a large energy saving benefit from the supply of surface water for domestic uses in urban and rural areas, which were earlier dependent on limited poor quality groundwater available in hard rock aquifers. While transporting water from a distant source like Narmada involves huge costs for the infrastructure, it is important to note that the water available from local sources is of poor quality, rendering it unfit for drinking and cooking. Since drinking water is a basic survival need, the cost of transporting water will have to be compared against the cost of producing freshwater from saline groundwater in the area, to test its cost effectiveness. In such a case, the investment for transporting water to Saurashtra and Kutch regions can be justified. Over and above, the local groundwater-based water supply sources are not perennial and dry up during summer months, and as a result, the region's urban local bodies have to go for drilling new wells, and diverting water from surface sources, at huge costs. In addition to the direct costs, there are also opportunity costs, which are by and large negligible.

9

Environmental Externalities of the Sardar Sarovar Project

With economic growth, India' electricity demand is growing. The per capita electricity consumption in the country increased from 1,204 kWh to 4,816 kWh during 1970–1971 to 2010–2011. In aggregate terms, the electricity consumption in the country increased from 43,724 GWh during 1970–1971 to 694,392 GWh during 2010–2011, as per a recent estimate. This shows a compounded annual growth rate of 6.98 percent. Of the total electricity sales in 2010–2011, the industry sector accounted for 38.6 percent, followed by the domestic (23.8 percent), agriculture (19.6 percent), and commercial sectors (9.89 percent). The electricity consumption in the domestic and agriculture sectors increased at a much faster rate compared to other sectors during 1970–1971 to 2010–2011, with CAGRs of 9.67 percent and 8.61 percent, respectively (NSO, 2012).

A major portion of the electricity consumed in the country is from coal-based thermal power. Out of the total 844,846 GWh of electricity generated in the country, only 114,257 GWh comes from hydropower plants. The energy mix in India is heavily skewed toward coal-based thermal power as a result of the introduction of water in the command areas and the associated changes in farming activities. During the past four decades, the country has seen a gradual shift toward coal-based thermal power from a good mix of thermal power and hydropower. Further, the thermal power comes from a nonrenewable natural resource.

While hydropower accounted for 45 percent of the total electricity generation in 1970–1971, today only 14.5 percent of the generation comes from hydropower. Increase in power generation through fossil fuel has enormous implications for greenhouse gas emissions (NSO, 2012). One of the strategies to reduce the greenhouse gas emissions from power generation is to change the energy mix, by increasing the share of hydropower and nonconventional renewable energy sources in India's energy basket.

The SSP, once completed, would add 1,450 MW to the India's installed capacity of power. It has been generating hydropower since 2004–2005. In lieu of the fact that the country is facing major power shortage, hydropower generation would reduce the dependence on thermal power to meet this shortage, thereby reducing the environmental damage caused by carbon emission.

The rising income level would increase the demand for environmental management services (Rosegrant et al., 1999). With impressive economic growth and rise in per capita income, the demand for environmental services has gone up in Gujarat, particularly in the urban areas. One of them was the desire to reestablish the flow regimes in the rivers in central and north Gujarat whose natural flows have been drastically altered due to diversions or impounds upstream for agriculture and other uses. Some of these rivers also had great values for the cultural life of the people of the regions they pass through. The Government of Gujarat took a major initiative to release a part of the unutilized flows from the Sardar Sarovar reservoir into these rivers to improve their ecological health and maintain healthy ecosystem. This was expected to add to the ecological benefits from the project. On the other hand, the introduction of exogenous water for irrigation and the consequent changes in farming practices can cause undesirable effects on the environment, particularly problems of waterlogging and salinity in the groundwater.

In this chapter, quantification of the economic value of the positive externality induced by hydropower generation on the environment or the benefit due to reduced carbon emission and the ecological benefit accrued through the release of water into the environmentally stressed rivers of Gujarat is done. The potential environmental hazards that are likely to be caused by the rise in groundwater table or the increase in groundwater salinity are also examined at length.

Benefits of Reduction in Carbon Emission

Gujarat is a state facing acute shortage of power, with the demand from industry, agriculture, and domestic uses exceeding the internal generation. Lack of generation from hydropower means greater dependence on thermal power. Therefore, every unit of power produced from hydroelectric projects reduced the dependence on thermal power by an equal amount.

We have obtained data on the actual power generation from both the river bed power house and canal bed power house of the SSP, for the period from 2004–2005 to 2010–2011. The values (m. units) are presented in a graphical form in Figure 9.1. From these we have estimated the mean value of annual power generation from the SSP to be 3,436 million units. We would like to mention here that since the entire power house with all the turbines became fully operational only in 2006–2007, only the values for five years since then were considered for calculations. We have estimated the positive externality of hydropower, which replaces thermal power generation, using equation (14) in Chapter 3. The average welfare benefit accrued from power generation was estimated to be ₹1,614 million per annum.

Figure 9.1:
Power Generation from the SSP 2005–2011 (Million Units)

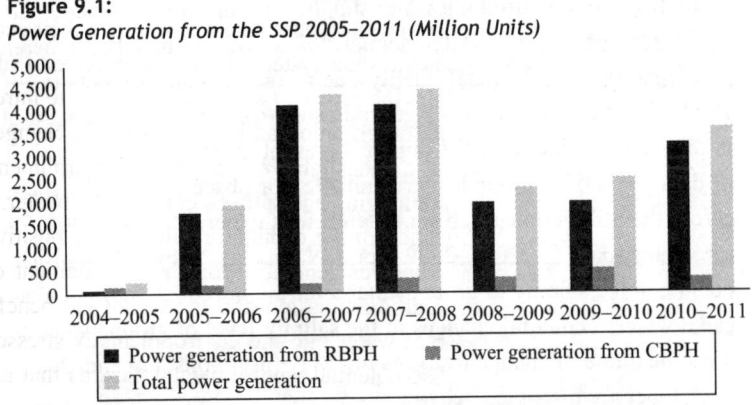

Source: Narmada Control Authority (http://www.nca.gov.in/news_index.htm).

Revenue Accrued from Sale of Electricity

Now, it is a well-known fact that hydropower is the cheapest form of electricity that can be produced from various renewable and nonrenewable sources, including the nonconventional sources of energy. As mentioned earlier, a look at the energy mix of the state's power generation shows that the largest chunk of the electricity produced in the state comes from thermal power. Hydropower is the second largest contributor to the state's energy pool. If we assume an average price of ₹5 per unit of electricity sold in Gujarat and an average electricity generation per annum of 3,600 million units from the SSP, then the total revenue earned from the sale of electricity generated in the SSP is ₹18 billion per annum.

Environmental Impacts of the Introduction of Narmada Water for Irrigation

Changes in the Salinity of Groundwater and Soils

One of the remarkable positive impacts of the introduction of surface water from canals for irrigation was on groundwater. As we have discussed in Chapter 5 on the positive externalities of canal irrigation on groundwater-based drinking water supplies, groundwater in many parts of Gujarat, particularly in the semiarid and arid regions, had inherent problems of poor chemical quality due to the presence of salts (GEC, 1997). Increased recharge of groundwater from freshwater can result in the natural dilution of the minerals present in groundwater. Analysis of data on TDS of groundwater available, for observation wells spread across the SSP command, from SSNNL for the period from 1996–1998 (pre-Narmada) to 2004–2009 (post-Narmada) showed significant decline in the salinity of groundwater after the introduction of Narmada canal water. Temporal changes in the salinity level of groundwater and the annual rate of change from May to May (pre-monsoon) and October to October are discussed below.

Pre-monsoon Change

In six out of the seven districts, the TDS values of groundwater in the designated command area either continued to decline (Ahmedabad, Baroda, and Bharuch) or the rate of the increase in TDS reduced (Banaskantha, Kheda, and Mehsana) during 2004–2009. The only exception was Surendranagar district where the trend got reversed from drop in TDS during the pre-Narmada period to rise in TDS post-Narmada. In the Narmada command area, the observed value of the salinity of groundwater (in May 2009) ranged from a lowest of 666 ppm to the highest of 3,256 ppm in the east (including Savali, Vadodara, Dabhoi, and Sankheda) and 380 ppm to 4,232 ppm in the west (including Jambusar, Amod, and Wagar).

Post-monsoon Change

In the post-Narmada period (2004–2009), the average TDS values of groundwater either continued to decline (Ahmedabad, Bharuch, Mehsana, and Surendranagar) or started showing a positive reversal of the trend (Baroda and Kheda). In the Narmada command area, the salinity of groundwater (in October 2009) ranged from a lowest of 770 ppm to the highest of 2,984 ppm in the east (including Savali, Vadodara, Dabhoi, and Sankheda talukas) and 370 ppm to 5,512 ppm in the west (including Jambusar, Amod, and Wagar).

As regards soil salinity, in the selected districts, farmers also confirmed decrease in the soil salinity. There is a decline of soil salinity in the command area, particularly in central Gujarat covering Ahmedabad, Gandhinagar, and Mehsana districts. There is a reduction in the inherent salinity of groundwater in these areas due to its dilution, occurring as a result of the additional recharge from canal water which is free from salts. Further, there is the leaching of salts from the alluvial sandy and sandy loam soils covering most parts of these districts, due to application of canal water during both kharif and winter seasons. The good vertical drainage of soils facilitated this process. The reduction in soil salinity is evident from all the districts in the command area; farmers were growing vegetables, which was not a feasible proposition prior to the introduction of Narmada waters due to high salinity in soils and well water.

Impact on Waterlogging

Problems of waterlogging are generally found to occur in areas that receive continuous recharge from irrigation return flows, but with less outflows.[1] Such situations can lead to positive water balance and "groundwater buildup." The primary survey in the SSP command area showed that there is widespread use of groundwater for irrigating crops. At the time of the field survey, no incidence of waterlogging (depth of the groundwater level equal to or less than 1.5 m) was seen in any of the selected districts (Ahmedabad, Bharuch, Mehsana, Narmada, Panchmahals, Surendranagar, and Vadodara). The lowest depth of the regional groundwater table (2.74 m) was recorded from the command areas coming under Narmada district.

Though water levels have been rising throughout the command area (except Kheda) owing to the import of surface water, no undesirable consequences were noticed. This is because of the fact that most of these areas have been facing problem of groundwater overexploitation (with the exception of Bharuch, Kheda, and Narmada districts) for more than three decades. Groundwater abstraction is far exceeding annual recharge in these areas. There has been a dramatic change in the groundwater environment since the time of the preparation of the plan document of the SSP in 1989, manifested by sharp decline in the water level throughout north and central Gujarat and Saurashtra, and dewatering of shallow aquifers.

Further, the present trend of rising water levels is not likely to continue in future due to the following reasons. First, the average depth of watering from the SSP is likely to stabilize in the coming years, as the second phase of the command area gets ready to receive water. The maximum average delta would be 18 inches (after the system losses) once the water distribution and delivery network for the whole command area (of 1.8 million ha) is in place, as against 22 inches of water being received by the farmers as of now. Second, once the dosage of surface water reduces to the design delta of 18 inches, the groundwater withdrawal in the command area would increase both in the south and north, as canal

---

[1] The outflows can be in the form of groundwater pumping or lateral flows due to groundwater flow gradients.

water along cannot meet the water requirement of the crops grown in two seasons, further reducing the possibility of water table build up. It is also important to note that the SSP will not be able to fully meet the demand for irrigation water in the regions where the canal network is extended, and the balance has to come from other sources, particularly groundwater. Around 40 percent of the cultivable land is under irrigation in Gujarat, and a larger share of the arable land in the semiarid central and north Gujarat and Saurashtra is unirrigated. Therefore, groundwater use would continue to be on rise over time, alone with the increasing population pressure. This would reduce the future possibility of waterlogging due to Narmada canal water. Nevertheless, there are poorly drained soils in certain areas within the SSP command. Such areas would require careful treatment in order to avoid waterlogging and salinity.

Changes in Fertilizer and Pesticide Application

As we have seen in socioeconomic impacts of the report, overall, there is an increase in the use of fertilizers after the introduction of canal water. The current average fertilizer application by farmers in the command area is much higher than the average figures available for the districts as a whole. For instance, between 1989–1999 and 2006–2007, average fertilizer consumption (Nitrogen, Phosphorous, and Potassium) per ha of cropped area at the district level increased from 61.4 kg to 92.7 kg in Ahmedabad, 133.7 kg to 146 kg in Kheda, 76.6 kg to 158 kg in Vadodara, 44.4 kg to 101 kg in Panchmahals, and 41.9 kg to 110 kg in Bharuch district.

Whereas for the canal irrigators, the average fertilizer use for the irrigated winter crops is 100 kg per bigha in Bharuch, 155 kg per bigha in Narmada, 135 kg per bigha in Panchmahals, and 161 kg per bigha in Vadodara, for irrigated kharif crops, the fertilizer use is 146 kg per bigha in Bharuch, 159 kg per bigha in Narmada, and 145 kg per bigha in both Panchmahals and Vadodara. In the case of canal lift irrigators, the average fertilizer use for the winter crops is 120 kg per bigha in Ahmedabad, 157 kg per bigha in Mehsana, and 82 kg per bigha in Surendranagar. For kharif crops, it is 96 kg per bigha in Ahmedabad, 163 kg per bigha in Mehsana, and 87 kg per bigha in Surendranagar. In most of the districts,

Figure 9.2:
Average Pesticide Consumption per Hectare of Net Sown Area in Gujarat

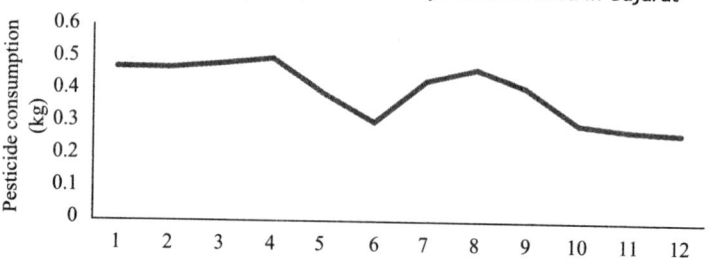

Source: Compendium of Environment Statistics India, Central Statistical Office, Ministry of Statistics and Programme Implementation, Government of India.

farmers are currently applying higher dosage of fertilizers when compared to the pre-Narmada period when it was about 84 kg per bigha in Ahmedabad, 105 kg per bigha in Mehsana, and 107 in Panchmahals. However, use of fertilizer application was found to have reduced slightly in the districts of Narmada and Vadodara post-Narmada.

On the other hand, the pesticide use per unit of cropped area has reduced substantially post-introduction of Narmada canal water. From the average of 0.47 kg per ha of the net cropped area in 1995–1996, it has come down to 0.27 kg per ha of the net cropped area in 2006–2007. Figure 9.2 shows the trend in pesticide consumption over the years in the state.

Ecological Benefits of Introducing Narmada Water and Its Economics

River basins which experience excessive diversion of water for various uses in comparison to the renewable water resources, or those which face environmental water scarcity (Kumar, 2010), water demand for ecosystem services could be very significant (Yang and Zehnder, 2005). All the river basins in alluvial north and central Gujarat are highly degraded with the following manifestations: overappropriation of surface water resulting in drying up of rivers for large stretches (Desai and Joshi, 2008;

Kumar *et al.*, 2001; Kumar, 2002), groundwater overdraft with lowering of water levels and drying up of shallow aquifer (CGWB, 1998), the disposal of effluents into rivers and the lack of flows to assimilate pollution (Kumar *et al.*, 2001), and drying up of ponds due to the absence of groundwater outflows and high rates of sedimentation.

Ecosystem water demand is defined as the water required for sustaining eco-environmental systems. In the specific context of the alluvial plains of north and central Gujarat region, it will have the following components: (1) the minimum in-stream ecosystem water use, (2) water use in urban lake and river environment, (3) water for reviving the wetlands (ponds), (4) water for groundwater replenishment, (5) water for sediment transport and estuaries, and (6) water use for pollution dilution. As pointed out by Pan and Zhang (2001), the ecosystem water demand can be set at two levels. The first level is the water required for preventing ecosystems from further deterioration, or for maintaining the status-quo, or the minimum water requirement for ecosystems. The second level is the water required for not only preventing ecosystems from deterioration but also progressively restoring degraded systems through water and soil conservation, afforestation, sediment and salt balance, assurance of base flows in rivers, urban ecosystem construction, and compensating the groundwater overdraft accumulated over the years. The second level of water demand is regarded as the amount required for maintaining healthy ecosystems (Pan and Zhang, 2001; Yang and Zehnder, 2005).

From the above, it is quite clear that there cannot be any simple and standard norms on the amount of water required by a basin for restoration of its ecosystem and that it would entirely depend on the extent of degradation of the basin ecosystem, climate, and hydrology of the basin; geo-hydrological environment; soils and physiography; and finally what level of ecosystem demands need to be met from the provision of flows. But from the outcomes of water release into the rivers of north and central Gujarat, it is clear that the second level of ecosystem water demands is met partly in some areas. For instance, there has been marked improvement in groundwater levels in the region, indicating that the total recharge from river flow and rainfall in the region is sufficient to cover the accumulated groundwater deficit and provide additional recharge. The trunks of the rivers, which used to get accumulated with effluent, are now free from the effluents and instead freshwater flow is available in all the rivers.

A survey was undertaken in the months of March–April, 2011, in the area located on the banks of two rivers which received releases from the Narmada canal to meet the ecological needs. They are Sabarmati River and Rupen River. While Sabarmati, which originates from *Ambe Bhawani* in the hills of Udaipur district of Rajasthan, Rupen originates in Mehsana and drains into the saline tract of Sami-Harij in the same district. Apart from understanding people's perception of the benefits from the induced river flows, the survey covered two aspects: (1) the amount of money people are willing to pay for the environmental management services provided by the increased flows in the river (Contingent Valuation method) and (2) the expenditure they have to incur to enjoy similar services in the absence of the same. The results vis-à-vis the perceived benefits are presented in Table 9.1. As Table 9.1 indicates, the major benefit perceived by people is recreational in both the locations. While 18 out of the 20 people surveyed in Ahmedabad felt that new plants and animals have started appearing in the area because of changing flow regimes in Sabarmati River and Mehsana, no such changes in flora and fauna were perceived by the local people, who live on the banks of Rupen River. Further, 6 out of the 20 people in Ahmedabad reported to

Table 9.1:
Characteristics of Induced River Flow and the Perceived Benefits

Characteristics of	Ahmedabad	Mehsana
Name of the river	Sabarmati	Rupen
When did you last see flow in the river?	2011	2009
How long did the flow last (in months)?	3.33	6.28
Year of introduction of Narmada water in the river?	2005	2005
Maximum flow-occurring season	Monsoon	Monsoon
Appearance of water	Turbid, colored	Turbid = 41 Colored = 20
Do people do fishing in the river? Commercial/ recreational	Recreational	Recreational = 49
Do people collect any other plants or living organisms from the river for consumption?	Six out of 20	None
Presence of new plants and animals in the area	18 out of 20	No = 49

Source: Based on a primary survey.

be dependent on the river for the collection of plants and other living organisms for consumption.

The contingent valuation procedure did not show abundant willingness on the part of the families to pay for the environmental services provided by the induced flow in the rivers. The services included fishing for recreation, swimming, boating, and seeing clean river water. A major reason for this is that water was introduced long ago, and people believe that releasing water from the Sardar Sarovar reservoir into the rivers is the best use of water, which yields benefits such as groundwater recharge and ecological health of the rivers without any opportunity costs or extra cost, as the command areas in central and north Gujarat are still not ready for receiving water, and that the agency would continue the practice of water release into the rivers without any public pressure. Therefore, the maximum amount, which the surveyed families have expressed their willingness to pay to the agency, was ₹50 per annum.

The only way to estimate the economic value of the environmental services from the induced river flows was the travel cost method, wherein the cost of traveling to the nearest site where similar facilities are available is worked out. The results of the travel cost method are presented in Table 9.2. In the case of Ahmedabad, the minimum distance to be covered was estimated to be 1.6 km, with the values ranging from 0.5 km to 2.5 km. The average travel cost per family per trip was ₹11.4,

Table 9.2:
Travel Cost Calculations

Particulars	Ahmedabad	Mehsana
Distance from the present site to site 1:	1.59	
Mode of transportation to that site?	Two wheeler	–
Approximate round trip cost to visit that site? (₹)	11.4	–
Duration of your visit per trip, on an average? (min)	18.0	–
How many times in a year, you would ideally visit that location (in the absence of the present facility)?	4.5	–
Approximate annual cost of travel to the site (₹) = (c) × (e)	47.6	
Amount of income foregone for the visit per trip (₹)	14.4	
Amount of income foregone per annum	65.0	
Total travel cost (₹) = (f) + (h)	112.6	

Source: Author.

with the mode of transport ranging from a bicycle to a scooter to a motor bike. The average number of trips per family was estimated to be 4.5 per year, and the total travel cost per annum, including the income foregone during the trips is estimated to be ₹112.6 per family, with a total duration of travel to the site being 18 minutes.

We can safely assume that in Mehsana too, an average family would be willing to pay nearly the same amount, that is, ₹113 per annum. So, the figure of ₹50, which the families are willing to pay as the annual fee for the environmental management services, seems reasonable in lieu of the fact that the agency had been releasing water into the rivers, without any demand from the communities. Though the figure might appear quite small, it is because some other ecological benefits and externalities of water release into the rivers such as improved groundwater recharge, rise in groundwater levels, and groundwater quality amelioration are not considered in valuing the environmental management services. In fact, they induce several positive externalities such as energy saving benefits in groundwater pumping for both irrigation and domestic water supply, and expansion in the well-irrigated area, but are separately estimated.

***

The indirect economic impact of clean energy production through hydropower from the project is a major positive externality of the SSP. It was estimated to be ₹1.61 billion per annum for a total estimated average energy production of 3,436 million units per annum. As regards the potential negative externalities on the environment, there was no incidence of waterlogging reported in the command area of the SSP, though water levels had risen in most parts of the command. This is because most parts of the command area had experienced groundwater overdraft for the last two to three decades.

As regards possible negative externalities on the society, the possibility of waterlogging appears to be less because of the following reasons: (1) the average depth of watering from the SSP is likely to come down in the coming years, as the second phase of the command area becomes ready; (2) once the dosage of surface water reduces, the groundwater withdrawal in the command area would increase both in the south and north, as canal water alone cannot meet the water requirement of crops grown in two seasons, further reducing the possibility of water table build up; and (3) as the SSP will not be able to fully meet the demand for irrigation water in the regions where the canal network is extended, the deficit has to be met from groundwater, and the use of the same is likely to increase over time, with population growth. Even today, only 38–40 percent of the gross cropped area in the state is irrigated.

The findings of the study have major implications for planning for the power sector in India. The energy mix of the country today is heavily skewed toward thermal power, which uses fossil fuel with major adverse effects on the environment. India's electricity demand is growing exponentially (NSO, 2012). The future investments in fossil fuel-based power generation to meet its growing demand would be subjected to greater scrutiny, as India is committed to meeting emission standards for carbon. Also, there is a need for maintaining good hydrothermal power mix for operational flexibility in power supply. The total hydropower potential of the country was estimated to be 84,404 MW.[2] Nearly 39.4 percent of the country's hydropower potential comes from the NE, which is mainly in the Brahmaputra and Barak basins, and another 35.9 percent was from the northern region. Only 2 percent of this potential is currently tapped (Rao, 2006).Against this, in the country as a whole, nearly 27 percent of the hydropower potential is currently tapped.

Part of the reason for the poor utilization of the potential in the NE is the growing resistance from the environmental lobby and grassroots NGOs in the region to dam-building on the grounds of the displacement of local communities, forest submergence, and ecological destruction.[3] But contrary to the popular belief, the potential submergence and displacement due to impoundment by storage reservoirs would be much lower in the NE region as compared to the southern and western regions,

---

[2] The total potential was assessed by the survey of the Central Electricity Authority in 1978–1987 at 60 percent load factor, and considered 845 projects.

[3] The other reasons include the huge engineering challenge of building dams in a way it does not induce seismic activities, the excessively high silt load carried by rivers in the Brahmaputra basin and the changing course of Brahmaputra and its tributaries.

owing to the topographical conditions and characteristics of river flow (Rao, 2006). The large environmental benefits of replacing fossil fuel-based power with hydropower should motivate the water and power sector planners in the country to invest in hydropower projects with minimum submergence of the pristine forests of these regions, if proper institutional arrangements can be worked out for adequately compensating for the PAP, and their R&R.

10

The Socioeconomic Impacts on Displaced Population

The SSP is the most meticulously planned multipurpose megaproject in modern India, which has a major bearing on irrigation efficiency thereby economic viability of the project. But with the several controversies shrouding the SSP, this has been hardly recognized (Patel, 2001: pp. 316). Smooth execution of the project was hampered by delays in getting environmental clearances for raising the dam height. Fisher (2001) argued that part of the delay was because of the lack of vital information about the social impact on the affected people in the submergence area. They concern matters as vital as the nature of relocated sites, the compensation that would be paid to the oustees, the eligibility of land entitlement to encroachers, and the infrastructure and amenities which the oustees would be able to access in the relocated sites (Patel, 2011). Very few attempts were made by the government agencies to provide this crucial information to the involved parties.

In the past two decades, the Narmada valley has seen one of the six largest anti-dam movements in the world. The communities living in the valley, in the upstream of the proposed megaprojects, were mobilized by environmental groups, pioneered by NBA, to protest against dam-building which is expected to cause submergence of land and the forests around (Fisher 2001; Patel 2001): The most effective opposition came from groups whose concern was survival of local cultures and autonomy of communities (Esteva and Prakash 1992). The six megaprojects planned in the Narmada valley put together are estimated to submerge a total land area of 141,170 ha affecting 80,000 families (Narmada Valley Development Authority undertaking). It can be seen that 38 percent of the PAFs and 65 percent of the submergence areas are located in the upstream of the Indira Sagar Project.

The anti-dam activists termed the projects a "genocide unparalleled in human history," crying bitter on the potentially negative social and environmental consequences, particularly submergence of vast stretches of forest land and relocation of tens of thousands of people. The pro-dam people on the other hand had eulogized the proposed projects as "lifelines of Gujarat and MP."

As regards the SSP, there are still serious contentions about the extent of "balancing" or "offsetting" of the negative social and environmental impacts of dam-building by the achievement of developmental goals such as increased water supply, agricultural growth, power generation, etc. But as Fisher (2001) pointed out, no methodological breakthrough in cost–benefit analysis can resolve this issue due to the ideological positions taken by the contesting parties. Under such a circumstance, what is more important is to know the condition of the dam oustees who have been rehabilitated under the R&R programs. Such a stand is important because no amount of benefits to the communities for which the project had been planned and built, would justify the trauma which displacement causes to the PAP, in addition to the social, economic, and cultural losses. Scudder (2005) categorized the stress of dam-induced resettlement into physiological (health impacts), psychological (separation from community and anxiety over the future), and sociocultural (impacts on cultural identity and livelihood support patterns) dimensions. Analysis of these aspects is also important in the wake of the criticism that the cost of the SSP "is being paid by the people who are least able to afford it and who will benefit the least from the fruits of development."

The History of the Evolution of the R&R Policy of Gujarat

Although international aid for the SSP came from many sources, the most significant one was the World Bank assistance, which was to account for

15 percent of the total estimated cost of the project. In 1985, the World Bank decided to lend the three state governments of Gujarat, MP, and Rajasthan funds to finance both the dams and the canals. The financing represented international approval of the project and satisfaction with the preparedness of the state governments to address the probable undesirable social and environmental consequences of implementing a large water resource project of the size of Sardar Sarovar. But, as the events unfolded, the issue of the dam oustees was internationalized by a few influential NGOs such as NBA, which expressed serious concerns about the condition of the rehabilitated people and went on to argue that complete rehabilitation of the displaced communities is impossible. This came to symbolize as one of the most embarrassing and criticized projects of the World Bank and became the first project to be terminated by the Bank.

The internationalization of the issue of the dam oustees, and the subsequent appointment by the World Bank of an independent review mission headed by anthropologist, Dr Bradford Morse, whose report strongly recommended termination of Bank's financial assistance for project construction on various grounds, finally led to the withdrawal of the Bank from the project. It is important to note that the independent review by Morse and Berger (1992) was also subject to scathing criticism from scholars. Patel and Mehta (2011) argued on the basis of a detailed analysis of the independent review that it suffered from serious methodological flaws, which include conclusions that do not flow from facts, the omission of vital information that could lead to entirely new sets of conclusions than what the review team arrived at, and distortion of facts (Patel and Mehta, 2011, p. 377).

Gujarat did not have a policy for rehabilitation of the dam oustees prior to the SSP. The NWDT, while declaring the award, also established specific conditions regarding the R&R of the PAP. The tribunal award stipulates that:

> All project affected families would be re-established as communities with access to water, education, and health (Clause IV [1]) on a "land for land basis." Clause IV (6) states that "In no event, shall any areas in Madhya Pradesh and Maharashtra be submerged under the Sardar Sarovar project unless all payments of compensations and costs is made for acquisition of land and arrangements are made for rehabilitation."

As Patel (2011) noted, the tribunal award was a trendsetter in a sense that for the first time it acknowledged the need for offering "land for land" to agricultural families instead of paying cash compensation under the LAA of 1870.

Though the norms for R&R in the SSP were decided by the conditions set out in the NWDT award and the R&R policies of Gujarat, the norms had undergone several modifications to take the final shape. Today Gujarat sets out most liberal R&R policies in the country (Patel, 2011; Sah and Tomar, 2011). Historically, several protests and representations made in various forums by NGOs together with the tribal population living in the valley against the conservative norms proposed by the Government of Gujarat for rehabilitation, the R&R guidelines set by the World Bank for lending (Patel, 2011; Sah and Tomar, 2011), and the greater understanding amongst the academics in the region about the sociocultural background and the special needs of the tribal communities (Sah and Tomar, 2011), have resulted in the Government of Gujarat reconsidering the R&R norms initially formulated and make them more realistic for the PAP.

Since 1980, ARCH-Vahini, a Gujarat-based NGO working for the rights of tribal population in Gujarat fought single handedly for a just rehabilitation policy under the SSP (Patel and Mehta, 2011). One of the serious issues of contention between the NGO and the state government was of giving the right to the tribal people likely to be displaced, to pick up the land of their choice for resettlement. This was done with a view to give the right to the oustees to choose the group he/she wants to resettle with, and the latter in turn chooses the village and the land for resettlement (Patel, 2011). The R&R policy of the state of Gujarat finally came into effect in 1987, which accepted this demand from the NGO that represented the tribal people in negotiations with the state government.

The R&R norms adopted by Gujarat for the PAP rehabilitated in Gujarat are described in Table 10.1. In addition, the package included civic amenities,[1] a grant-in-aid, and subsidies for purchase of productive assets to the tune of ₹7,000 per PFA.

---

[1] The civic amenities are as follows: one primary school with three rooms for every 500 families; one children's park for every 500 families; one panchayat ghar (see note) for every 500 families; one dispensary for every 500 families;

Table 10.1:
Norms for R&R as per the NWDT Award and Norms Followed by the States

Category of PAF	NWDT Norm for Compensation	Gujarat's R&R Norm	Government of MP	Government of Maharashtra
Landowners	2 ha of agri. land and house plot	2 ha of agri. land, house plot, and ₹45,000 as assistance for construction of core house	2 ha of agri. land and house plot	2 ha of agri. land and house plot
Landless farm laborers	No agri. land. Only House plot	2 ha of agri. land, house plot, and ₹45,000 as assistance for construction of core house	No agri. land. House plot given with productive assets. ₹49,300 (SC/ST) and ₹33,150 for others	1 ha of agri. land and house plot
Encroacher	No agri. land. Only house plot	2 ha. of agri. land, house plot, and ₹45,000 as assistance for construction of core house	Allotment of agri. land of 1 ha or 2 ha subject to the size of encroachment; and house plot	2 ha of agri. land if the encroachment prior to March 31, 1978. For encroachments after this period, 1 ha of agri. land and house plot given

Source: SSPL, June 2012.

Though the construction of the proposed dam at Kevadia started, soon after the NWDT gave its award and had continued even after the World Bank discontinued its financial assistance for the project. The work had to be suspended on June 5, 1995, when the dam was constructed to a height of 81.5 m as NBA filed a writ petition in the Supreme Court expressing serious concerns about the quality of rehabilitation of the PAP.

---

one seed store for every 500 families; village pond for every 500 families; one drinking water well with trough for every 50 families; one tree platform for every 50 families; one religious place of worship for every 500 families; linking of colonies to the main road with by-roads of appropriate standards; and electrification, water supply, and sanitary arrangements (SSPY, 2012). [Note: Every village has a representative body called panchayat. The panchayat has five members. Panchayat house is the village-level governing office.]

The petition requested that no further submergence of the land or displacement of people living in the proposed reservoir area takes place. Clearly, the objective of the anti-dam activists, who joined hands with the environmental groups from around the world, was to internationalize the issue of displacement surrounding the SSP as a human rights issue. They argued that complete rehabilitation of the oustees would be impossible, and that the government should give up the construction of the project itself to protect the rights of the indigenous people in the Narmada valley (Fisher, 2001; Verghese, 2001; Patel, 2001). As pointed out by Patel (2001), this had a major impact distracting the state government from matters concerning R&R during those years, as the entire effort of the administrative machinery was in obtaining the stay on the writ vacated so as to resume the dam construction.

However, after examination of the cases presented by both the parties and an affidavit filed by the governments of Gujarat and Maharashtra that all the oustees have been rehabilitated and further arrangements have been made for the rehabilitation of those whose land would be submerged with increase in dam height by another 3.5 m, the Supreme Court in an interim order allowed resumption of the work till the height of 85 m. Since then, the Government of Gujarat had been obtaining environmental clearance for raising the height of the dam in 5 m stages. The height of the dam currently stands at 110 m, against the proposed height of 138.5 m.

The Status of Rehabilitation

In order to facilitate efficient implementation of R&R, a new institutional mechanism was created by the Government of Gujarat, named, Sardar Sarovar Punaradhivas Yojna (SSPY), with the following cells, viz., land cell, engineering cell, rehabilitation cell, agricultural cell, medical cell, education cell, and grievance redressal cell.

The SSPY recognizes the PAF as either a landowner, a major son of the landowner, landless agricultural laborer, encroachers, or co-sharers.

The SSP had one of the largest number of PAFs, who were displaced by reservoir submergence. It was estimated that a total of 46,658 families would be displaced by submergence of land in the water spread area of the reservoir. They are called the PAFs. a small proportion of these

Table 10.2:
Total Number of PAFs in Different States in the Upper Catchment of Reservoir

Name of the State	Village Affected			PAFs Likely to Be Resettled	PAFs Likely to Be Resettled in Gujarat	PAFs Already Resettled in Gujarat
	Fully	*Partially*	*Total*			
MP	1	191	192	37,691	5,517	5,517
Maharashtra	0	33	33	4,201	747	747
Gujarat	3	16	19	4,766	4,766	4,765
Total	4	240	244	46,658	11,030	11,029

Source: SSPY, June 2012.

families (11 percent) actually are from Gujarat, a majority of them are from MP, and a few from Maharashtra (747 in number). But as per the NWDT award, the number of PAFs to be settled in Gujarat (11,029) is much higher than those who were displaced from within the territory of the state (5,517). See Table 10.2.

The R&R settlements in Gujarat are spread across south Gujarat extending from Narmada district in south Gujarat, where the reservoir is located, to Ahmedabad district in central Gujarat. Of the total 11,029 PAFs, who are to be rehabilitated by Gujarat, 4,765 families are from Gujarat. They are rehabilitated in 110 sites, spread across seven districts of the state. But, 75 percent of them are in Vadodara district. While, the families from MP were rehabilitated in 108 sites, more than 50 percent are located in Vadodara district. Similarly, as regards the oustees from Maharashtra, 75 percent of the sites are located in Vadodara. In the case of Maharashtra oustees, majority of them are located in Vadodara (14 sites), and the rest are in Narmada (3 sites) and Surat (1 site). In total, 153 out of the total 236 R&R sites are in Vadodara. The second highest number of sites (32) is in Narmada district. See Table 10.3.

Of the total of 46,658 families which are likely to be affected by the project from the three states of MP, Gujarat, and Maharashtra, a total of 19,000 are expected to prefer Gujarat for resettlement. But a total of 11,030 families are to be resettled in Gujarat. Of these, 11029 families have already been rehabilitated. Further, of the total 10,925 families which were eligible for agricultural land, 10,925 already received land. The total land allocated to these families was 21,862, which means on average, each PFA got a total of 2.0 ha of land, which was the norm.

Table 10.3:

Resettlement Sites of Sardar Sarovar Dam Oustees in Gujarat

| District Name | No. of Sites for Resettlement of the Oustees | | | |
	Gujarat PAFs	MP PAFs	Maharashtra PAFs	Total
Vadodara	82	57	14	153
Narmada	27	2	3	32
Kheda	0	14	0	14
Bharuch	0	13	0	13
Ahmedabad	0	11	0	11
Panchmahals	1	10	0	11
Surat	0	1	1	2
Total no. of sites	110	108	18	236
Total no. of PAFs	4,765	5,517	747	11,029
No. of families per site (approximate)	43	51	42	47

Source: SSPY, June 2012.

Plots for house construction had also been allotted to 10,975 out of the total 11,029 families. Of the total 11,029 PFAs, 4,028 families had already been shifted to the rehabilitation sites, and in 9,269 cases, houses had been constructed with productive assets such as wells dug in 9,167 cases. Interestingly, insurance coverage has been provided to a total of 11,063 families, which is a little more than the total number of families rehabilitated.

The subsidy of ₹7,000 per PAF is given for purchase of productive assets. These assets include cows, electric motors, bullocks, bullock carts, oil engines, and agricultural implements. The detail of the productive assets purchased by the PAFs is given in Table 10.4. It is to noted here that the families which have shifted from their original habitat to their new habitat must have also brought along with them the assets owned by them. Another important factor which needs to be kept in mind while looking at these data is that out of the 11,029 PAFs, only 4,028 had shifted to a new habitat, though the land, house plot, etc., had already been allotted to most of these families. More than 9,174 families have already acquired the productive assets.

Table 10.5 shows the distribution of productive assets purchased by oustees of SSP books in the three states.

Table 10.4:
Status of Rehabilitation of Sardar Sarovar Dam Oustees

Particulars	Unit	Gujarat	Maharashtra	MP	Total
Total PAFs (Allocated)	No. of PAFs	4,766	747	5,517	11,030
(Resettled)	No. of PAFs	4,765	747	5,517	11,029
(Balance)	No. of PAFs	1	0	0	1
Agricultural land					
Eligible PAFs: (Allocated)	No. of PAFs	4,726	743	5,457	10,926
(Allotted)	No. of PAFs	4,725	743	5,457	10,925
(Balance)	No. of PAFs	1	0	0	1
Total land eligibility (Allocated)	(In ha)	9,464	1,486	10,914	21,864
(Allotted)	(In ha)	9,462	1,486	10,914	21,862
(Balance)	(In ha)	2	0	0	2
House plots					
(Allocated)	No. of PAFs	4,765	747	5,464	10,976
(Allotted)	No. of PAFs	4,764	747	5,464	10,975
(Balance)	No. of PAFs	1	0	0	1
Action Plan/Time frame for R&R of balance PAFs	No. of PAFs	0	0	0	0
PAFs shifted to R&R site from submergence	No. of PAFs	2,975	605	448	4,028
Transportation grant	No. of PAFs	0	0	0	0
Payment of subsistence allowance	No. of PAFs	4,715	628	5,079	10,422
Payment of rehabilitation grant	No. of PAFs	4,260	234	156	4,650
Payment of ex-gratia	No. of PAFs	4,330	0	0	4,330
Core house constructed	No. of PAFs	4,116	546	4,607	9,269
Productive assets provided	No. of PAFs	4,613	685	3,876	9,174
Employment provided	No. of PAFs	393	11	13	417
Insurance cover	No. of PAFs	4,760	748	5,555	11,063

Source: SSPY, June 2012.

Table 10.5:
Distribution of the Productive Assets amongst the Oustees from the
Three States

Name of Productive Assets	Gujarat	Maharashtra	MP	Total
Bullocks (Nos.)	5,878	1,098	6,895	13,871
Bullocks cart (Nos.)	343	74	126	543
Cow (Nos.)	99	39	54	192
Electric motor (Nos.)	15	0	1	16
Oil engine (Nos.)	4	0	16	20
Agri. equipments	1,947	390	1,177	3,514

Source: SSPY, June 2012.

The Socioeconomic Impact of Relocation on the Oustees

Ideally, the relocation of PAFs would bring in a spate of changes in their lives, when it is accompanied by changes in the economic environment, wherein the traditional barter economy of tribal societies is replaced by the market economy. In such situations, the oustees may need support to visualize and get adjusted to competitive markets and have resources and information to become active participants in technologically dynamic agriculture in the host villages (Sah and Tomar, 2011). Their success in adjusting to the psychological trauma and new socioeconomic and institutional environments by the PAFs would depend on how much of this support is enabled by the institutional interventions for rehabilitation.

Ownership of Farm Land

Shah and Tomar (2011) reviewed the changes in the socioeconomic conditions of the PAFs after relocation. The review used analysis carried out by Sah (1994) on the basis of data generated in 1996, when nearly 4,557 families were fully rehabilitated and 96 percent of them received agricultural land. They found that relocation not only increased the land assets of the PAP, but also changed the way this land is owned and operated by them. In the submerged villages, a significant proportion of the households (43 percent) used to practice joint operations on the agricultural land. In sharp contrast to this, in the new settlements, 92 percent of the holdings were operated by individuals. In the new villages, all the cultivated land was privately owned. The average holding was 5.7 acres, with a total operated holding of 5.9 acres.[2]

Encroachment was rampant in the submerged villages. For every acre of the land cultivated in the submerged village, more than an acre was cultivated illegally in the forest land. As a result, while the average operational holding of the HH was only 4.2 acres, the land area actually operated was 9.2 acres. This was mainly because a significant share (32 percent) of the PAFs did not own land and were cultivating on illegally occupied forest land. Also, those who owned land in the area were also encroaching upon the forest land for cultivation.

The survey carried out for the present study in one of the settlements of Bhil tribes in Bharuch district showed that the average landholding per household is now 5.12 acres, which showed a decline from 8.48 acres in the presettlement situation. But the average size of the family also came down sharply from nine persons to just six persons. Thus, the average landholding per capita has actually not declined significantly. Over and above, the extent of encroachment during the native settlement was very high and was estimated to be 91 percent (7.69 acres) per household. Hence, the actual amount of land owned by the households was much less (0.88 acres or 0.15 acre per capita). But, today, the entire land is privately owned.

Changes in the Farming Enterprise

Significant changes in the farming enterprise were also visible after the relocation to the new area. The traditional low (economic) valued crops and coarse cereals were replaced by fine cereals and cash crops. While around 42 percent of the area was under coarse cereals (jowar), the proportion of its area came down to 22 percent in the new villages.

---

[2] The survey for the pre-displacement scenario used data from 14 out of the 19 submergence villages, and the relocation scenario used data from 120 new sites (Sah, 1993, 1996).

Further, only 2 percent of the cropped area was under paddy in the submerged villages; in the new settlement, around 14 percent of the area was under paddy. Moreover, cotton was grown in 12 percent of the area in new villages while this crop was not grown at all in the submerged villages. Percentage area under chick pea has increased after relocation. Nearly 4 percent of the area was under fodder in the new site. These data suggest that farming of the tribal communities had become market oriented after resettlement. Overall, while 98 percent of the area was under cereals and pulses in the submerged villages, in the new sites, the proportion came down to 83 percent, and a significant share of this (45 percent) is fine cereals.

The most recent survey carried out as part of the current study also corroborated with the findings of earlier studies. It showed that the number of farmers growing crops such as wheat and maize had increased, whereas fewer farmers grow jowar, a traditional cereal crop used by the tribes in their native villages. Their proportion came down from 86 percent to 17 percent (Table 10.6). Irrigation of cereal crops has also become quite extensive in the new settlement. It is as high as 87 percent for millets, 79 percent for maize, 48 percent for paddy, and 27 percent for wheat, while none of the farmer used to irrigate his crops in their old settlements.

This had impacted the crop yields also. The yield of all the cereals was reported to have gone up remarkably (Table 10.7). The most significant increase was in the case of wheat, followed by paddy. However, this improvement cannot be attributed to irrigation alone. The farmers have also started using fertilizers, as a result of exposure to modern farming practices. While only two households reported the use of fertilizers in the native settlement, 29 of the 30 families surveyed reported using chemical fertilizers presently. The average expenditure on fertilizer per household is ₹6,033 per annum.

The survey also showed changes in livestock holding of families after R&R (Table 10.8). The most notable change was in the ownership of small ruminants. While, on average, every family in the old settlement had a little more than five goats/sheep, in the new settlements, they were not reported to own these small ruminants. As regards cows and buffaloes, the number of cattle has gone up from 4.05 to 4.57, while that of

Table 10.6:
Changes in Crop Choices, Cropping Pattern, and Irrigation Pattern with Resettlement

Name of Crop	Number of Farmers Growing the Crop		Average Area (acres) under the Crop		Extent of Irrigation	
	Presettlement	New Settlement	Presettlement	New Settlement	Presettlement	New Settlement
Paddy	6	6	1.91	1.50	0.0	47.80
Wheat	1	7	2.00	1.50	0.0	26.70
Maize	25	29	2.14	1.47	0.0	78.90
Jowar	26	5	2.06	0.90	0.0	55.60
Millets	20	22	1.84	0.99	0.0	86.80
Other cereals	11	23	1.42	1.65	0.0	85.60

Source: Authors' own analysis based on primary survey, 2013.

Table 10.7:
Change in the Yield of Cereal Crops

Name of the Crop	Average Crop Yield (Quintals per Acre)	
	Presettlement	*New Settlement*
Paddy	1.0	5.0
Wheat	1.0	8.0
Maize	2.0	4.0
Jowar	1.0	2.0
Other millets	1.0	2.0
Other cereals	1.0	4.0

Source: Authors' own analysis using primary data, 2013.

Table 10.8:
Change in Animal Holding of Households with Resettlement

Livestock type	Presettlement	New Settlement
Indigenous cow		
In milk	1.16	1.00
Dry	1.36	1.32
Calf	1.53	2.25
Total	4.05	4.57
Total per capita	0.45	0.75
Buffalo		
In milk	1.16	1.00
Dry	1.00	1.10
Calf	1.83	1.50
Total	4.00	3.60
Total per capita	0.44	0.60
Bullock	1.58	1.85
Bullock per capita	0.17	0.32
Goat	4.17	—
Sheep	1.00	—
Small ruminants per capita		

Source: Authors' own analysis based on primary data, 2013.

buffaloes came down from 4 to 3.6. But it is also important to take into consideration the fact that the size of the household had drastically come down from 9 to 6. So, the average size of livestock holding per capita had increased substantially after resettlement.

Technology Adoption

Survey carried out by Sah (1994) to compare and contrast the farmers in the new settlements and the farmers in the host villages showed that the landholdings of the sample PAFs is larger than that of the farmers in the host villages. While all the sample farmers belonging to the PAFs were belonging to medium farmer category, the farmers in host villages were small and medium. But the gross income per hectare of area cropped was less for the sample farmers as compared to those in the host villages (₹1,953 per acre against ₹2,464 per acre). The difference was attributed to sharp differences in input use between farmers in the host villages and those from the PAFs, and the adoption of new crop technologies such as high yielding varieties. As Sah (1994) noted, they also had poor access to institutional credit and adopted self-financing of the current production approach out of the fear of falling victim to the local money lenders and interlocked markets.

Given the fact that only a small percentage of the PAFs had irrigation facilities, they were found to be engaged in water purchase. While in some locations, they were buying water from the farmers belonging to Patel community, in some other locations, the PAFs were part of groundwater irrigation cooperative societies they themselves formed. In fact, 90 percent of the irrigation in the new settlements of the PAFs was from private water markets, though most of them were unhappy with the irrigation services provided to them by the Patel farmers (Sah and Tomar, 2011). Nevertheless, those who are dependent on water markets were found to have earlier better agricultural income than those who are dependent on rain-fed farming.

Access to Drinking Water Sources

In the native settlement, the families used to depend on natural water bodies such as rivers/streams and springs for their drinking and other domestic uses. Only 10 percent of the households used to get water from public bore wells, and the distance traversed to collect water was also remarkable (nearly 1.0 km on average). But, in the rehabilitated settlements, the households are provided with adequate number of public water sources, with the result that nearly 87 percent of the households depend on hand pumps for the domestic water supplies (Table 10.9). The average distance to the source has also reduced drastically to a mere 28 m from a little more than 1.0 km. This is a huge benefit, in terms of saving in time spent on water collection. As regards the quality of water, nearly 56 percent of the households reported about the poor quality of water used for domestic purpose, in their native settlement, and as high as 90 percent reported the water to be of good quality.

Access to Health Facilities

The survey carried out in one of the settlements in Bharuch district showed remarkable improvements in access to health facilities, including government hospitals and private clinics. In the old settlement, every household took the services of traditional medical practitioners from their community. Only 6.7 percent of the households used the public health clinics and another 6.7 percent went to government hospitals for medical care (Table 10.10). But, as Table 10.9 shows, today, more than three-fourth of the households depend on the government hospitals for medicare. Ninety percent of the households also depend on the primary health

Table 10.9:
Change in Degree of Access to Formal Water Supply Sources with Resettlement

Source	*% HHs depending on for Domestic Water Supplies*		*% HH depending on Livestock Drinking*	
	Presettlement	*New settlement*	*Presettlement*	*New Settlement*
Tanks	0.0	3.3	10.0	3.3
Public bore wells	10.0	6.7		6.7
Hand pumps		86.7		9.0
Piped water supply		3.3		
Tankers and others				
Rivers/stream/springs	90.0		90.0	
Purchase	6.7	0.0		0.0
Average distance of source from household (in m)				

Source: Authors' own analysis using primary data, 2013.

Table 10.10:
Change in Access to Medicare with Resettlement

Type of Medical Service	Presettlement	New settlement
Traditional medical practitioners	100.0	66.7
PHC	6.7	90.0
Registered medical practitioner		6.7
Government hospitals	6.7	76.7
Private hospitals		20.0
Percentage of Prenatal Deaths		
	Presettlement	New settlement
	3.3	0.0

Source: Authors' own analysis using primary data, 2013.

centers (PHCs). Twenty percent of the households also take the services of private hospitals. However, in spite of all the improved medical facilities, nearly two-thirds of the surveyed households still depend on the traditional medical practitioners in the new settlement.

Coping Strategies During Stress

Irrigation has a major role in stabilizing agricultural production and making farming enterprise less vulnerable to vagaries of monsoon. Only 10 percent of the gross cropped area in the new site was already under irrigation in 2011. But, since a large percentage of the sites are under SSP command (80 out of the 126 sites), they were likely to receive Narmada water for irrigation. A survey carried out in 1995 showed changes in agricultural productivity, calorie intake of the PAFs over time. The survey covered 1991–1992, 1992–1993, 1993–1994, and 1994–1995. It showed that farm productivity in economic terms rose substantially from a lowest of ₹3,003 in 1991–1992 to ₹10,048 in 1993–1994 in the new sites, while it was a little higher than the 1991–1992 figures in the submerged villages. But in 1994–1995, the outputs dropped drastically to ₹3,309 due to droughts. The low farm productivity in 1991–1992 and 1994–1995 were due to droughts. The number of farmers who reported crop losses was as high as 40 percent during 1991–1992 and 30 percent during 1994–1994. Now the impact of the reduction in farm outputs on the calorie intake of the households depends on how the households are able to cope with the stresses.

The data showed that the proportion of the HHs reporting low calorie intake (< 2,400 kcal) increased to 72 percent in 1991–1992 from 53 percent in the submerged villages, but it went on dropping to become 33 percent in 1994–1995. This was in spite of the decline in the value of agricultural outputs during that year. One reason for this, as pointed out by Sah and Tomar (2011) was that the households might have used their surplus production during the previous years to tide over the crisis during the year.[3] Another reason could be that only 30 percent of the farmers reported crop losses during the drought of 1994–1995, though the average income was only a little above the 1991–1992 figures.

While it is understood that the ability to cope with droughts would depend on the socioeconomic characteristic of the family, the attempt to identify the variables which explain the change in the calorie intake was made by Sah (1994). The analysis found that out of the 10 variables chosen, five variables, viz., casual labor, agricultural income, total number of family members, cattle sale, and migration had high correlation with per capita calorie intake. The analysis found that factors such as milk production, service, loan taken, and the gross cropped area did not have much impact on the calorie intake (Sah, 1994, as cited in Shah and Tomar, 2011). Multivariate analysis using these five explanatory variables found that these five factors explain the changes in calorie intake only to an extent of 16 percent. As Shah and Tomar (2011) noted, a low R2 value indicated that some of the most crucial explanatory variables were not considered in the analysis. Some of them could be past saving, current borrowings and grain stock.

An important issue relating to calorie intake which has been missed in the analysis by scholars is that calorie intake is not always a reflection of economic distress. Over the past several years, the calorie intake in the rural areas had declined in spite of reduction in poverty rates. Particularly, with the use of modern transportation families, the increased mechanization of farming, and the change in lifestyles with lesser hardship in domestic work, the calorie intake requirement of family members could have reduced significantly. In this case, the tribal households in

---

[3] Sah and Tomar, 2011, based on Sah, 1994.

their original habitat were more dependent on human and animal labor for farming operations, and were used to walk long distances for fetching water and firewood. In the new settlements, such hardships are less due to access to modern farming equipments, better access to water supply sources, and lower dependence on firewood from forest land.

In the tribal societies, the communities had their own mechanisms of absorbing shocks caused by economic distress during droughts resulting from the failure of crops. These mechanisms are mainly in the form of social cohesion and social and cultural ingenuity, which are indigenous institutions. Such indigenous institutions are absent in the new settlements of the host villages.

More importantly, the natural capital which they could bank on in their habitats, during times of distress, no longer exists. This had affected their modes of collection of food. The tribes used to collect a wide variety of food (roots, fruits, flowers) in addition to producing food in their own farms. The bush fruits and meat of animals and birds from hunting become a major source of survival during times of distress. Nearly 50 percent of the households reported to have collected forest produce as food even under normal circumstances. Nearly 7 percent reported hunting also as a source of food. But the percentage of households using forest produce had drastically come down in the new settlement (Table 10.11).

The tribal households have, however, devised new coping strategies to overcome the impacts of droughts in their new settlements. For instance, outmigration of some of the family members in search of casual work had increased during the drought years. Reduction in the use of purchased inputs, reducing expenditure and changing the consumption

Table 10.11:
Changes in Modes of Food Collection

Mode of Food Production	Presettlement	New Settlement
Hunting	6.7	3.3
Collection from forest/nature	50.0	3.3
Animal keeping	96.0	56.7
Produce from own farm	100.0	96.0
Purchase from local market	96.0	100.0

Source: Authors' own analysis using primary data, 2013.

pattern to less expensive items are other coping strategies adopted by the affected families.

Overall Satisfaction with Life in the New Settlement

For many of the families surveyed, it has been nearly three decades since they were relocated to the new settlement. While a large proportion of the respondents were having bad memories about their life in the old settlement (50 percent) in terms of happiness, a large majority (86.6 percent) feel good in the new settlement (Table 10.12). One-sixth of the respondents did not respond to the specific queries regarding the degree of satisfaction. As Table 10.12 indicates, among them nearly 30 percent feel very happy (that is, 23.3 percent of the total respondents). Only 13 percent still feel unhappy with the living in the new area. Nevertheless, nearly 97 percent of the households felt that they were getting adequate support, cooperation, and acceptance in the host village.

Findings and Conclusions

The results emerging from the analysis can be summarized as follows: (1) With a liberal R&R policy adopted by the Government of Gujarat, the displaced families including the encroachers who possessed land in the submerged villages, the majors in the family, apart from the agricultural landowners, and agricultural laborers got 5 acres of land; (2) with greater access to own land, the livelihoods of the affected families

Table 10.12:
Change in Overall Level of Satisfaction with Life with Resettlement

Degree of Happiness	Presettlement	New Settlement
Very good	3.3	23.3
Good	16.7	63.3
Satisfactory	13.3	3.3
Bad	50.0	13.3

Source: Authors' own analysis using primary data.

have become heavily dependent on agriculture in the new settlements; (3) the overall agricultural income of the affected families have increased after relocation to the new sites owing to the increased yield of the crops grown and shift toward to commercial crops; (4) the overall agricultural income of the displaced families in the new settlement, however, is lower than that of the farmers of the host villages, as their adoption of improved crop and input use technologies was much lower than that of the farmers in the host village; (5) in lieu of the adoption of crops that require higher dosage of inputs including water, the income from farming has become more vulnerable to rainfall fluctuations. The access to drinking water supply and public health had also improved remarkably after resettlement in terms of quality and easiness in accessing the facilities. For the communities, the overall degree of happiness with life in the new settlement is much higher than that in the native settlement.

Way Forward

Given the emerging water and power scenarios in the states of Gujarat and MP, there cannot be much doubt over the relevance of "large-scale interventions" in ensuring water and livelihood security and more sustainable growth of vast areas of Gujarat and MP. In fact, the "need" for water and energy has never been the source of contention in projects such as the SSP in the Narmada valley. But as William Fisher puts it, "the needs, no matter how great, do not serve as a justification for a specific solution" (Fisher, 2001). In fact, the World Bank now acknowledges that a particular dam should be selected for construction only after a careful analysis of all other options that include not only other dams, but also other means of meeting the same need (Fisher, 2001, p. 310). Whether such options were never rigorously analyzed in the context of the Narmada valley before deciding for megaprojects is a matter of debate. But the fact remains that those who critiqued dam-building never mooted such convincing alternatives to Sardar Sarovar.

Having said that, the *locus standi* that reducing the size of the affected population and the social impacts thereof are of utmost importance remains unchallenged. This is strengthened by the emerging international standards and blueprints for good practices in the design, selection, and implementation of energy and water projects, and advanced tools and techniques now available for making ecologically sound designs of water and energy projects (Fisher, 2001, p. 306; Paranjape and Joy, 1995). The opposition against the Indira Sagar Project, which is crucial for integrated water development in the Narmada valley, is the biggest, in view of the large submergence and the number of people to be dislocated.

Generating vital information about the social and environmental impacts of the Indira Sagar Project—downstream impacts, health impacts, and impact on the affected people—would be crucial: a weakness area for the SSP. Lack of detailed information about the number of people likely to be adversely affected, the nature and degree of impact, and economic losses resulting from it could hinder the planning to mitigate impacts as it happened in the case of the SSP. ARCH-Vahini highlighted the problem of proper rehabilitation and resettlement of the oustees hindering the process of smooth implementation of the SSP and the need for high precision execution in a complex project like the SSP (Patel, 2001, p. 326). Since, the anti-dam movements (especially NBA) still hold on to their long-standing view that the social and cultural costs of the dam cannot be compensated and full rehabilitation is impossible, the Narmada Control Authority needs to generate accurate information about the impacts and also perform strict monitoring to ensure proper rehabilitation of the oustees.

11

Maximizing Future Benefits and Minimizing Negative Impacts from the Sardar Sarovar Project

The standard approach to CAD in India is based on the premise that surface irrigation would lead to excessive seepage and return flows from canals would lead to waterlogging and salinity problems in the absence of sufficient control on water delivery and farmers shifting to water-intensive crops. An associated concern was that water distribution would be highly inequitable across the hydraulic system, with more water being made available to the farmers at the head reaches and very little or no water reaching the tail-end farmers. As a consequence, suggestions for CAD schemes focused on the lining of canals, building of drainage networks, promotion of conjunctive use of surface and groundwater, and creation of farmer organizations for equitable distribution of water among the farmers in the command area, along with many other irrigation management functions (Kumar and Bassi, 2011).

Such views emanating from the discussion on the efficient management of irrigation were rationale at a time when well irrigation was not in vogue, groundwater was underdeveloped, pump irrigation was also too expensive owing to scarcity of electricity in rural areas and the poor advent of energized pump sets, water was not as much a scarce resource as it is today in many regions, and the infrastructure and technologies to access water was a major issue.

Today, problems are different. Groundwater overexploitation has become one of the greatest environmental problems facing the state, with depletion and quality deterioration problems. They are more severe in central and north Gujarat (Kumar, 2007), which are likely to get the least benefit from the SSP in terms of the area irrigated (Raj, 1991). The agriculture sector has become one of the largest consumers of electricity, which it uses for pumping groundwater, and the state's revenue losses in the form of electricity subsidy for groundwater pumping amounts to nearly ₹40 billion annually, which increases every year with a decline in groundwater levels across the state. A major chunk of this goes to central and north Gujarat, where water tables are deep (Kumar and Bassi, 2011). Under such situations, as pointed out by Allen *et al.* (1998) and Chakravorty and Umetsu (2003), seepage from canals cannot be treated as a waste or a negative externality. Instead, it becomes a valuable resource, which can be reused through wells.

On the other hand, due to excessive diversion of water, rivers of north and central Gujarat are experiencing severe environmental water stress (Kumar, 2002). Groundwater quality deterioration affects thousands of villages in north and central Gujarat, which are exposed to poor quality groundwater for drinking, with a myriad of water-related health problems such as kidney stone and fluorosis (IRMA/UNICEF, 2001).

While on-farm water management and equitable water distribution would still be great challenges to be addressed in irrigation management and need to be taken up, there are far greater challenges of maximizing the economic, social, and environmental benefits from the project in view of the positive externalities that large water resource projects like the SSP can induce on the society. As the analysis presented in the previous chapters suggested, these benefits are many.[1] They can outweigh

---

[1] The SSP makes positive contribution to improving groundwater recharge in overexploited areas, sustaining well irrigation and reducing well failures (Ranade and Kumar, 2004; Shah and Kumar, 2008); reduction in the cost of pumping groundwater (Shah and Kumar, 2008; Vyas, 2001); reducing carbon emissions through generation of clean energy (Joshi and Kapadia, 2009); improvement in groundwater quality and the resultant improvement in the water supply situation in villages and urban areas with positive health benefits; improving the health

some of the direct benefits of irrigation. But, there is little appreciation of these externalities (Kay *et al.*, 1997; Shah and Kumar, 2008).

But, not all benefits can be realized uniformly across the hydraulic system. Groundwater recharge is an important benefit, in an area where the aquifer is mined. Ecological benefits of augmented flow regime would be significant if rivers experience environmental water stresses. Benefits of groundwater quality improvement will be significant in areas where natural quality is poor or moderate, but not extremely poor. The SSP command cuts across several agro-climatic subregions (Raj, 1991) and also several geo-hydrologic environments (Kumar, 2002). The south Gujarat part of the command receives high rainfall (1,000–1,200 mm) and has subhumid climate and shallow aquifers, whereas the north and central Gujarat parts of the command have low to medium rainfall, semiarid to arid climate, and heavily exploited deep alluvial aquifers. Saurashtra has low to medium rainfall and shallow, hard rock aquifers. Rivers in south Gujarat are perennial, while those in central and north Gujarat and Saurashtra experience environmental water stress (IRMA/UNICEF, 2001; Kumar, 2002).

The extent to which we can realize these benefits depends on how judicious is the allocation of water from the SSP across regions. It is quite clear that the usual interventions taken up under a CAD program would not help and, on the contrary, can often be counterproductive if implemented on a blanket basis. This means that the objectives of CAD programs and criteria for assessing them should be broadened to inter-nalize the changing socio-ecological realities of the region. The follow-ing section would illustrate why the conventional approach is not going to bring about optimum benefits and a unique strategy is needed in the case of the SSP, to maximize the economic, social, and environmental benefits from the project, which include both direct and indirect benefits (Kumar and Bassi, 2011).

---

of riverine ecosystem by the release of water into the rivers or north and central Gujarat (Shah and Kumar, 2008; Desai and Joshi, 2008); and lowering of cereal prices (Perry, 2001a; Shah and Kumar, 2008) and gain to the state exchequer in the form of reduced electricity subsidy for the farm sector.

The Challenges Facing SSP

In Chapter 4, we have discussed the sharp regional imbalances in water resource endowment and water demand in Gujarat. The SSP involves transfer of water from the naturally water-rich south Gujarat, which has low water demands, to three other regions of the state, which are naturally water-scarce, viz., north Gujarat, Saurashtra, and Kutch, but have excessively high water demands. Hence, it is a rarest of the rare hydraulic infrastructure projects in the world having the capacity to level off the regional imbalances in water availability and the demand in a large geographical area. It can change the agrarian economies of all the four regions of Gujarat, viz., south Gujarat, north and central Gujarat, Saurashtra, and Kutch. Though the project has already become the major driver of Gujarat's agricultural growth (Kumar *et al.*, 2010b), faces some major challenges.

SSNNL has not been charging for irrigation water on volumetric basis, in the area where the commands are receiving water. The farmers, therefore, have strong incentive to go for crops that are highly water-intensive such as banana, even in the naturally water-scarce regions of north and central Gujarat, and also use water excessively. Such a trend has been seen in well irrigation in Gujarat (Kumar, 2005), Uttar Pradesh (UP), and south Bihar (Kumar, 2009) under flat rate pricing of electricity wherein farmers are confronted with zero marginal cost of using electricity and, therefore, groundwater. SSNNL needs to create infrastructure and put institutional mechanisms in place to make measurement and volumetric pricing of irrigation water a reality. Earlier analysis had shown that the farmers in south Gujarat who grow highly water-intensive banana and sugarcane incur much lower cost for a unit volume of irrigation water as compared to those who grow low-water-consuming crops due to the inefficient (crop area based) pricing in which the irrigation of high-water-consuming crops attract higher charge per unit area (Kumar and Singh, 2001).

Illegal lifting of water from canals by farmers who are both inside and outside the designated command area is emerging as a major challenge for the SSNNL officials. The fact that the SSP envisages a total of 66,000 km long canals (Desai and Joshi, 2008) poses an entirely new challenge to the irrigation engineers of the state. These regions have also been facing severe depletion of groundwater resources over the past several decades, and people in this region feel that getting more water from the SSP is their legitimate right. Certain shocking realities with regard to regional agricultural trade enhance the legitimacy of these claims (Kumar and Bassi, 2011). Analysis done as far back as 2004 showed that north Gujarat, in spite of being absolutely water-scarce, exports milk to other parts of the country, including regions which are water-abundant, the water content of which was estimated to be 1,100 MCM per annum (Singh, 2004). North and central Gujarat, though water-scarce, have more arable land and can produce surplus agricultural output, if irrigation water is provided. South Gujarat, though water-rich, has less amount of arable land (Kumar and Singh, 2005).

A big challenge which SSNNL is facing today in making water available to the farmers of north Gujarat is in completing the construction of the network canals. Managing funds for constructing the canals is a major challenge, while the agency had developed innovative ways of making land acquisition procedure smooth. Lining of canals takes a major chunk of the construction cost. Hence, there are two conflicting interests. One is to provide greater volumes of water for meeting irrigation needs in central and north Gujarat, which are agriculturally prosperous, but facing water resource crisis. The other is to reduce the cost of building water distribution infrastructure.

How to Maximize the Returns from Irrigation?

From the foregoing analysis, it is clear that to maximize the economic, social, and environmental returns from the project calls for more judicious allocation of water from the SSP amongst different regions to maximize the benefits of irrigation. Greater allocation of water to central and north Gujarat and Saurashtra, which are water-scarce, would produce incremental economic returns, over what it would otherwise generate if that water is used up in water-rich south Gujarat districts, as illustrated by studies undertaken in western Punjab and eastern UP in recent past which showed that the marginal return from irrigation is higher in water-scarce regions as compared to water-rich regions (Kumar *et al.*, 2008). Similar argument was made by the renowned irrigation economist, B. D. Dhawan, that marginal returns from irrigation would be high in low rainfall regions and low in high rainfall region (Dhawan, 1990). Also, greater allocation of water, which can result in enhanced groundwater recharge in north and central Gujarat, would bring about high social returns, through improved sustainability of well irrigation, reduced energy cost for pumping groundwater owing to the rise in groundwater levels, improvement in the quality of groundwater, and ecological benefits of improved flows in environmentally water-stressed rivers of north and central Gujarat. But, this has to be done without any changes in the legitimate (volumetric) allocations of the farmers in different regions, as per the delta decided for each agro-climatic subregion during the design phase.

Technical Approach

Water-scarce Areas of North Gujarat

In the initial years, the irrigation dosage from the SSP can be much higher than the delta determined (22 inches) by the Narmada Planning Group. This is possible, without changing the allocations earlier decided for each agro-climatic subregion, in view of the fact that .nany of the large water projects planned in the MP part of the basin in the upstream of the Sardar Sarovar Dam, which would eventually help utilize all the dependable flows in the basin (Kumar, 2010) are yet to be taken up (Ranade and Kumar, 2004). The water, which is released from the upstream reservoirs during the rainy season and which is in excess of the reservoir storage capacity of the SSP, can be diverted for irrigation through the Narmada main canal. This can be done at almost zero marginal cost as the infrastructure for conveying the water will already be in place, and what needs to be done is to increase in the duration of water supply from the canal system (Kumar and Bassi, 2011).

During these years, the maximum proportion of water carried through the tertiary canals in the rainy season and irrigation water applied in the field should be allowed to percolate down the unsaturated zone of north

Gujarat's alluvial aquifers by keeping the canals unlined (Kumar and Bassi, 2011) and choosing highly water-intensive crops. By virtue of the high moisture content of the soil due to the rainy season and unlined canals, the percolation would be high. Nevertheless, it is understood that those canals which run on filling will have to be lined with bricks for preventing breaches and stabilizing the embankments (Kumar and Bassi, 2011).

This would help raise the groundwater table (Ranade and Kumar, 2004). Research studies undertaken in the Murray basin of Australia and the Indus basin in Pakistan, both having arid climate and deep water table conditions, showed substantial rise in water table conditions after the introduction of canal water for irrigation. Such changes happen due to reduced pumping of groundwater with the introduction of canal water, gradual increase in moisture storage in the unsaturated zone, which increases the unsaturated hydraulic conductivity of the soil media, and the increase in moisture pressure (hydraulic) gradient of the soil (Watt, 2008). While the first factor reduces or keeps the vertical distance for the movement of recharge water, the second and third factors increase the rate of the vertical movement of soil water.[2]

Irrigation during the rainy season would increase the speed of the vertical movement of water (return flow) through the unsaturated zone by virtue of the high hydraulic gradient of soil moisture and higher hydraulic conductivity of soils in the unsaturated zone owing to a higher degree of saturation and low depth to groundwater. It will also increase the return flow portion of the water applied in the field due to a high degree of saturation of the root zone soils. This would accelerate the recharge process. Since the opportunity cost of diverting the water will be insignificant, the benefits arising from the improved groundwater recharge, the groundwater quality remediation, the improved quality of soils, and the rise in water levels could be treated as the additional economic benefits (positive externalities). The positive externalities include: the expansion in irrigated area by wells and, therefore, the well-irrigated production; the reduced investment for well deepening by farmers; and

---

[2] Based on Richards Equation and van Ganuchten Equation as cited in Watt, 2008, pp. 101–102.

the reduced cost of supplying water through alternate sources to villages and towns/cities affected by groundwater quality deterioration.

During the winter season, when the storage from the reservoir would be released for irrigation, the farmers can take up crops that give higher returns per unit volume of water, rather than those which are water-intensive and which give higher recharge per unit area of land irrigated. Nevertheless, it is important that larger area is covered by canal irrigation during winter, as this would reduce the dependence on groundwater for irrigation in the region, thereby reducing the imbalance faster. For this, the new commands does not have to be developed through the construction of canal networks. Instead, water from the distributaries can be diverted to areas outside the designated command area for farmers to irrigate their land.

Once groundwater resources are built up in the overexploited area, the canals can be lined depending on the groundwater-level fluctuations in the area. It is, however, unlikely that the water level rise would cause waterlogging and salinity problems. The reason is that the total agricultural water demand in the region cannot be fully met from canal water, and the farmers from the region would continue to depend on groundwater to meet their irrigation needs. The need for lining will not come so much from the fact that it is important for preventing waterlogging and salinity, but from the fact that seepage would result in the wastage of costly water. This is because the economic value of the benefits accrued from return flows would reduce gradually as the groundwater situation in the region improves. But, such a decision to carry out lining should be based on empirical studies on the actual rate of seepage from canals (Kumar and Bassi, 2011). Research has shown that over the years, the infiltration losses from unlined canals stop permanently as the water table below the canal bed level rises due to continuous seepage and percolation, saturating the entire underlying strata (Watt, 2008).

Water-rich Areas of South and Central Gujarat

In the water-rich areas of south and central Gujarat, which are underlain by shallow and phreatic alluvial aquifers, and shallow hard rock aquifers, the amount of water to be released for irrigation water supply will have to be limited. The reason is that excessive irrigation or irrigation of water-intensive crops such as paddy and sugarcane would lead to excessive return flows and finally the rise in groundwater levels. Since the groundwater balance in the region is still positive, this can create problems of waterlogging and salinity in the command areas, similar to what happened in the Mahi command. If irrigation dosage from canals is limited, farmers could depend on groundwater in the area to supplement it. This would increase the opportunities for allocating greater quantities of water for north and central Gujarat, which receive lower quantum of rainfall, and experience higher aridity and groundwater depletion.

There is a strong economic rationale behind transferring a part of the entitlement of south Gujarat to north and central Gujarat. As studies show, the surplus value product from a unit volume of irrigation water is much higher in a water-scarce region as compared to a water-rich region. This is because the farmers in water-scarce regions allocate their water more efficiently to different crops, from a pure economic viewpoint, by putting more area under crops that give higher water productivity in economic terms and improving the physical efficiency of water use. Also, the impacts of water used in irrigated agriculture are much higher on livelihood (Kumar *et al.*, 2008). Secondly, every unit of water transferred to north and central Gujarat for agriculture produces much large social, ecological, and environmental externalities, with great values realized in economic terms. Such positive externalities are either absent or are much less in south Gujarat.

The technical strategy for this would be lining the delivery canal systems, in addition to the distribution system. This would save the precious water, which otherwise would seep and percolate to shallow aquifers, creating waterlogging and salinity problems.

Promotion of the Use of Micro-irrigation Systems in the Command

One of the ways to maximize the efficiency of the utilization of water stored in the Sardar Sarovar reservoir is to improve the efficiency of its use in crop production. Unlike what is being proposed for the monsoon season, the water utilization strategy for winter has to be different as the inflows into the reservoir would be minimal. Here, the strategy should be to minimize the amount of water diverted and depleted.

MI systems can bring benefits in every agro-climatic condition in the SSP command, in terms of raising crop yields and getting better quality produce. But, the reason for adopting MI will be different in different environments. In south Gujarat, the purpose of using MI should be to reduce the irrigation water requirements for crops. It should also be to reduce the deep percolation which augments the shallow groundwater in the area, thereby preventing waterlogging problems. In north and central Gujarat, where water table is deep, the use of MI systems should be to reduce the non-beneficial, non-recoverable deep percolation into the unsaturated zone, and non-beneficial evaporation from land not covered by canopy (known as "non-beneficial consumptive use").

But, the MI systems are not amenable to canal systems due to the mismatch in water release schedules followed in canal systems and the irrigation schedules required for the optimum performance of the MI systems. For using canal water for irrigation through the MI systems, the following additional infrastructure needs to be built in the farmers' fields: (1) an intermediate storage system and (2) a pumping device for pressurizing the water (Kumar *et al.*, 2008). SSNNL has already set up 12 pilot projects on the use of drips and sprinklers in different locations within the SSP command.

But such systems would be viable only in areas where the land is available in plenty. The economic viability of MI systems would be better in areas where water saving is going to be maximum. Here, the real water saving comes from reduction in non-beneficial, consumptive use. Higher reduction in non-beneficial consumptive portion of irrigation water applied would occur when aridity is high, spacing between plants is large, depth to groundwater table is large, and underground formations are saline.

The Institutional Approach

The institutional approach to improving canal irrigation management in India had primarily focused on local institutional development, to enable farmer participation in the management of irrigation systems (Brewer *et al.*, 1999). Market instruments such as volumetric pricing of canal water

(Kumar and Singh, 2001), supply rationing (Kumar and Amarasinghe, 2009), and enforcement of volumetric water rights (Kumar and Singh, 2001) which are powerful tools for bringing about improvements in water use efficiency and effective demand management (Kumar and Singh, 2001; Kumar, 2010; Rosegrant and Schleyer, 1994, 1996) are not used (Kumar and van Dam, 2009). The experiments in institutional alternatives had looked at various models, ranging from those which are aimed at managing the irrigation affairs at the level of tertiary systems (Chambers, 1988; Merrey, 1996) to those which are meant for managing the entire hydraulic system starting from the main system to the tertiary system (Brewer *et al.*, 1999). However, the management functions envisaged for these local institutions included operation and maintenance of the system, water distribution, recovery of water charges, and conflict resolution. There is no dearth of examples from India of water users' associations enforcing water allocation norms amongst member farmers through the measurement of volumetric water delivery from the inlet of the tertiary system, that is, chak.

Irrigation Water Pricing

Kumar and others had argued that volumetric pricing can be an efficient tool for water allocation in canal irrigation (Kumar and van Dam, 2009; Kumar, 2010) though many scholars have argued that there are theoretical and practical issues in using volumetric pricing as a tool for irrigation water demand regulation (Perry, 2001b; de Fraiture and Perry, 2002). In the absence of a positive marginal cost of using canal water under the pricing structure that is based on the crop area, the farmers in canal commands have no special incentive to use water efficiently. In fact, an earlier analysis carried out in the context of south Gujarat had shown that the implicit volumetric price for many water-intensive crops, such as sugarcane, banana, and paddy, is much higher as compared to many low-water-consuming crops, such as bajra and jowar (Kumar and Singh, 2001). Further, since there are no restrictions on the volume of water delivered owing to water abundance, the farmers in the canal command areas of south Gujarat are found to be extensively growing certain water-intensive crops that have high land-use efficiency such as banana, sugarcane, and paddy (IRMA/UNICEF, 2001).

Volumetric pricing of canal water would motivate farmers to use water efficiently for irrigating the crops, as it induces the positive marginal cost of using water. But they might still grow crops that are highly water-intensive, using water efficiently (Kumar and Singh, 2001; Kumar, 2010). The reason is that they would be interested in maximizing the returns from a unit area of land, which is possible just by using more water under the present condition. This is because there are no effectively enforced restrictions on the volume of water that farmers can use (Kumar and Singh, 2001; Kumar and van Dam, 2009). It is important that farmers in central and north Gujarat use more water during the kharif season when the opportunity cost of using the same would be low, but the recharge benefits are relatively high. Therefore, it is important that the unit price of water is kept low for the water released during the kharif season.

Fixing Volumetric Water Entitlements and Water Trading

After Frederick (1992) establishing privately owned, property rights that are tradable, it is critical to establish conditions under which individuals will have opportunities and incentives to develop and use the resource efficiently, or transfer it to more efficient uses. If water entitlements of canal irrigators are fixed in volumetric terms, and if they are made tradable, regional water markets would emerge. The selling price of canal water is likely to go up in the market as demands are high. The demand for water is likely to come from naturally water-scarce regions of central and north Gujarat due to two reasons: (1) The farmers there would be able to generate higher return from every unit of water used in farming and (2) the area irrigated by the SSP would be much smaller in comparison to the total arable land and the potential of aquifers in the region to supply water to irrigate the remaining area is increasingly becoming poor. The water entitlement for the farmers in the command area can be fixed in such a way that it creates artificial water shortage. In reality, this would anyway happen after the completion of the project under the Rotational Water Supply Scheme (RWSS) proposed for the project command, as the volume of water for each chak is decided on the basis of the irrigable area and crop water requirement for the main irrigation season, that is, winter.

Well-defined tradable rights formalize and secure the existing water rights held by water users, induce water users to consider the full opportunity cost of water, and provide incentives for water users to internalize and reduce many of the negative externalities inherent in irrigation (Rosegrant and Binswanger, 1994; Kay *et al.*, 1997). If markets are efficient, the price at which water is traded would reflect the opportunity cost of using water (Howe *et al.*, 1986; Frederick, 1992). The price which north Gujarat farmers would be willing to pay for water would be the market price of water. Since groundwater markets already exist in the region, the price at which water is traded between well owners and water buyers can provide the price signals. Since the farmers in south Gujarat get low net returns from a unit volume of water used in irrigation for the dominant crops of the region such as banana, paddy, and sugarcane, they would have a great incentive to shift to water-efficient crops, or use it efficiently or sell it at the market price and earn income (Frederick, 1992; Kumar and Singh, 2001). Since the choice of crops that give high returns per unit volume of water (water productivity in economic terms) is limited in south Gujarat, farmers would be willing to transfer a share of their entitlement to central and north Gujarat and earn income. Given the high demand for water in north and central Gujarat, there would be large-scale water trading. The incentive for this would be high, as farmers do not have to invest in the infrastructure required for water transfer (Kumar and Bassi, 2011).

The Enabling Environment for Institution Building

If water pricing has to be introduced, measuring the volume of water used by farmers is a prerequisite. This has to be done at the level of the inlet of the tertiary system, and subsequently, the water delivery at the field inlet of the individual farmers will have to be monitored for the duration of water supply. There are some notable examples of volumetric measurement of water delivery by WUAs at the inlet of the tertiary system. They are Waghad and Mula irrigation projects in Maharashtra. On the other hand, if volumetric water entitlements have to be enforced, the measurement of the volume of water used by the farmers has to be done for protecting the rights of individual water users and to make the markets work perfect. This would also help monitor the physical transfer of water entitlements across regions and check its legitimacy. For the first time in India, canal automation project is being implemented in a large project like the SSP. Since the hydraulic heads in the canal system would be continuously monitored and controlled, the measurement of the volume of water delivered from each control point can be measured.

Since it would be impossible to work out water transfer agreements between hundreds of thousands of individual farmers located in two different regions, intermediate agencies will have to be created at various levels in the hydraulic system hierarchy—from the main canal to the branch and distributory canals. Transparency and accountability will have to be maintained by agencies that would be engaged in bulk water transfer at various levels in the hydraulic system. This would call for making the various levels in the hydraulic system independent, which means measurements of water delivery at various control points within the hydraulic system. This can also provide the basis of negotiation of water transfer agreements between parties, and payments for sale and purchase of water entitlements.

It is imperative that institutions will have to be created at various levels in the hydraulic system. SSNNL had long ago proclaimed that irrigation water in the SSP command will be supplied only to irrigation cooperatives or water users' associations and not to individual farmers. Hence, already there is an enabling policy environment in Gujarat. The need for such institutions would be quite felt by the farmers themselves unlike in the case of WUAs being promoted in other canal command areas. This is because there would be strong economic incentives to get organized for the economic benefits that can be derived from having access to irrigation water and specifically from water trading. In the past, the institution model promoted for farmer management of irrigation in the SSP had invited a lot of criticism from at least some quarters, arguing that it attracts huge transaction cost involved and there is a sheer lack of interest among farmers in participating in the irrigation management of the kind which was envisaged at the time of project planning (see Talati and Shah, 2004). But, if volumetric water pricing is practiced and tradable water entitlements are enforced, farmers and other groups at various levels in the hydraulic system hierarchy would show great willingness to form organizations to engage in water trading of different scales as such transactions would yield returns.

This can unlock the real values of canal irrigation through the SSP in Gujarat. Volumetric water transfer to the tune of 2,000 MCM to north Gujarat can earn farmers of south Gujarat and the intermediaries nearly ₹4 billion annually, if we assume that the farmers in north Gujarat pay ₹2 per m^3 of irrigation water purchased. This is quite reasonable in view of the fact that farmers from the region are paying more or less the same price for water from tube wells, which is of inferior quality, due to the presence of salts. More importantly, as Table 11.1, based on a study carried out way back in 2003, shows the net returns that farmers in the region get from the use of water for crop production are far higher than this. But the nature of institutions would be different. They should have the capability to take care of the upkeep of the system, measure water volumes being transferred through various control points in the hydraulic system, and prevent illegal lifting of water from the distribution and delivery canals.

What is even more important is that fact that there would be significant economic gains for the intermediate institutions engaged in water trading owing to the fact that the physical transfer of water (which is to be traded between the regions) would not take place from main system to the actual title owners of water in south Gujarat, once the trade

Table 11.1:
Water Productivity of Various Crops

Sr. No.	Name of Crop	Water Productivity for		
		Well Owners	*Water Buyers*	*Shareholders*
1	Wheat	3.83	4.01	5.61
2	Castor	6.99	9.19	
3	Mustard	5.21	3.88	5.10
4	Cumin	16.70	22.93	19.89
5	Jowar	4.01		
6	Bajra	3.67	2.04	
7	Cluster Bean	6.46		
8	Fennel	3.36		
9	Leafy vegetable			8.11

Source: Kumar, 2005.

agreements are negotiated between farmers in south Gujarat and those in central and north Gujarat. This would result in huge saving in water which would otherwise be lost in conveyance. The irrigation bureaucracy can give ownership of this water to the intermediate institutions, which would provide them strong monetary incentive to participate in water trading.

Finally, the monetary gains from water transactions should induce new incentives among the farmers below the minor outlets to form WUAs to carry out the larger irrigation management functions, which go beyond water transactions. Since such institutions will be self-driven rather than external agency-promoted, they are likely to sustain. That would also reduce the transaction cost of creating institutions.

***

The irrigation from the SSP produces multiple benefits. Given the fast changing water and energy environment in Gujarat, the economic value of these benefits can outweigh the direct benefits of irrigation, water supply, and hydropower generation. In order to maximize these benefits, irrigation water use in water-rich south Gujarat needs to be conservative, and more water needs to be allocated for central and north Gujarat. Enhanced water allocation for north and central Gujarat has to meet the needs of farmers, both within and outside the designated command area in that region. This can be done without changing the original water allocation plans and delta worked out for each agro-climatic subregion in the command, as unutilized water from the upper catchment of the Narmada basin would be available for many years for diversion from the Sardar Sarovar reservoir in addition to Gujarat's share of 9.75 MAF.

This would require institutional as well as technical interventions. Lining of delivery canals would be extremely important in south Gujarat, in order to reduce wastage of the water supplied and to prevent waterlogging and salinity in the long run. Contrary to this in north Gujarat, enforcement of tradable water entitlements would motivate farmers in the region to sell part of their water entitlements to farmers of north and central Gujarat. Subsidized volumetric prices for water diverted through Narmada canal during kharif season would encourage farmers of north Gujarat to use it for crops that are water-intensive crops. Trading of water can happen in bulk water transfer through the existing water conveyance infrastructure of the SSP. Institutions are likely to come up at different levels across the hydraulic system hierarchy to engage in water trading, as they would benefit immensely in monetary terms through such trading. But it is to be noted that unlike the WUAs which are promoted at the tertiary system and above, these institutions should have high levels of management skills so that they can take care of the upkeep of the distribution/delivery system, the measurement of volumetric water releases, and prevent illegal lifting of water from the distribution and delivery systems.

12

Conclusion

Globally, large water resource systems have an undisputable role in ensuring water, energy, and livelihood security, and therefore, in the social and economic wellbeing of the people in the regions they serve. According to the HDR (2006), only one in every five people in the developing world has access to an improved water source. Dirty water and poor sanitation account for vast majority of the 1.8 million child deaths each year (almost 5,000 every day) from diarrhea—making it the second largest cause of child mortality. In many of the poorest countries, only 25 percent of the poorest households have access to piped water in their homes, compared with 85 percent of the richest. Diseases and productivity losses linked to water and sanitation in developing countries amount to 2 percent of GDP, rising to 5 percent in Sub-Saharan Africa—more than the aid the region gets. These countries also face acute shortage of water for agricultural production, a mainstay of their economies. For instance, drought-prone Sub-Saharan Africa has the smallest proportion of its cultivated area (< 3 percent) under irrigation (HDR, 2006). Because of this, reduction in rainfall leads to decline in agricultural production, food insecurity, malnutrition, loss of employment opportunities, and an overall drop in economic growth in rural areas (Kumar, 2009).

These countries are characterized by poor public investments in creating water infrastructure projects and have very low per capita reservoir storage, which reduce their ability to deal with water-related disasters such as floods and droughts (Grey and Sadoff, 2007).The economic losses that water-related natural disasters caused in these regions due to the absence of adequate water infrastructure are huge (Kumar, 2009). For instance, in Ethiopia, deviation in per capita GDP from the normal values during the 20-year period from 1980–2000 correlated with the departure of annual rainfall from normal values (World Bank, 2006a). In Kenya, economic losses due to floods during 1997–1998 were to the tune of 11 percent of the national GDP, whereas that due to droughts during 1998–2000 were 16 percent of the GDP (World Bank, 2004, 2006b). In the Indian state of Gujarat, the value of the agricultural output dropped from ₹268.37 billion in 1998–1999 to ₹189.0 billion in 2000–2001 following the droughts in 1999 and 2000 (Kumar *et al.*, 2010b, Figure 1), yet, there are institutionalized attempts globally to downplay the role of large water resource systems in improving water security at the regional, local, and household level, thereby bringing about progress in human development and reducing hunger and poverty. Instead of this, the focus has shifted to building social infrastructure (Briscoe, 2011).

Sardar Sarovar is among the most complex water resource projects ever designed and built in independent India, to stabilize and sustain agricultural production and to provide water for humans and cattle in vast areas of a drought-prone province, with an aim to protect millions of rural livelihoods and reduce poverty. It is a multipurpose project aimed at providing irrigation, drinking water, industrial water supply, and hydropower benefits. It is also a well-designed project. But, the project got mired in controversies due to fierce protests from environmental and human rights activists, which questioned the viability of the project on social, economic, and environmental grounds. Because of this, there has been a major time lag between planning and implementation of the project. During this time, the water environment of Gujarat had dramatically changed. Groundwater depletion has become widespread, while the degree of groundwater development in Gujarat was very low during the 1980s. So is the change in overall socioeconomic conditions in different regions of the state. Water demands for irrigation, domestic uses, and industry increased exponentially, with consequent increase in the value of water. There has been a remarkable upsurge in energy demand in the state, particularly for industry and agriculture.

As a result, many of the benefits derived from the project today could not be anticipated at the planning stage. These benefits are far larger than the intended benefits (a summary of all direct and indirect benefits is provided in Table 12.1). For instance, the ecological benefit of the scale occurring today was never anticipated before. Instead, the project faced severe criticism for the potential ecological destruction it would bring about through submergence of catchments, water diversion, and water-logging in the irrigated command area. The recharge to groundwater through irrigation return flow was never considered as a benefit. Instead, it was expected to cause waterlogging and salinity, which are negative externalities. But with groundwater overexploitation, recharge has now become a boon, making well irrigation in the area more sustainable. Water level rise also results in the saving of precious electricity used for

Table 12.1:

Direct and Indirect Benefits from the Sardar Sarovar Narmada Project

Particulars	Economic	Socioeconomic	Environmental and Ecological Benefit
Canal Irrigation in command areas	Increase in returns and farm surplus from crop and dairy production	More reliable irrigation source Increase in cropped and irrigated area Increase in crop yield Increase in livestock holding Rise in wage employment for farm laborers Overall improvement in the economic condition	Groundwater recharge through return flows from irrigated fields and seepage from canal Increase in the availability of biomass Improvement in groundwater and soil salinity No observed incidence of waterlogging Reduced application of fertilizers and pesticides
Drinking water supplies	Reduction in expenditure on treatment of water-related diseases Reduction in expenditure on purchase of good quality water	Access to reliable and better quality water source Increase in the frequency of water supply and volumetric use of water Reduction in distance travelled and time spent on water collection	–

(Table 12.1 Contd)

(Table 12.1 Contd)

Particulars	Economic	Socioeconomic	Environmental and Ecological Benefit
Positive externality of canal irrigation on well irrigation in command areas	Reduction in the cost of groundwater pumping Increase in returns and farm surplus from crop and dairy production Saving in the cost of well deepening Energy saving benefits	Shift in cropping pattern toward water-intensive but high-valued crops Increase in irrigated area Increase in crop yield Increase in livestock holding Incremental income of wage laborers	Rise in groundwater level Reduced groundwater pumping Reduced carbon emissions Increase in the availability of biomass
Positive externality of irrigation and water supply on domestic water supply in command areas	Reduction in the cost of domestic water supply in command areas Reduced investment for well deepening Energy saving benefits	Changes in wage earning potential of women and men due to increased time availability	Rise in water levels in the wells Reduced carbon emissions
Power generation	Increase in revenue from sale of electricity	–	Production of clean energy Reduced carbon emissions
Release of Narmada canal water into other rivers	–	–	No major impact on hydrological regimes Appearance of new varieties of flora and fauna

Source: Author.

pumping groundwater. Generation of hydropower also resulted in the reduction of carbon footprint, which otherwise would have occurred if same amount of electricity was produced through the use of fossil fuel. These are all positive externalities on society or indirect benefits.

All these developments called for criteria to be used in the evaluation of the project benefits than those which were used in normal project benefit–cost analysis of such large water projects. This warranted designing innovative methodologies for quantifying these benefits.

The project benefits have reached many different regions of Gujarat, which are characterized by heterogeneity in physical (agro-climatic, geo-hydrological), socioeconomic, and cultural environments. The water environment in the alluvial, semiarid north Gujarat drastically differs from that in south Gujarat, which receives high rainfall. The hard rock region of Saurashtra is vastly different from both south Gujarat and north Gujarat. Parts of south Gujarat, which receive irrigation water from Narmada, are dominated by native tribes, whose agriculture is not as developed as in north Gujarat. Thus, the impact which exogenous water is likely to induce on the local water environment and the socioeconomic landscape is likely to be different from region to region. Hence, a comprehensive analysis of the project benefits was necessary.

During the past two decades, the Narmada valley has seen one of the six largest anti-dam movements in the world. The communities living in the valley, in the upstream of the proposed megaprojects, were mobilized by environmentalists, pioneered by NBA, to protest against dambuilding which is to cause submergence of their land and the forests around (Fisher, 2001; Patel, 2001), though the most effective opposition came from groups whose concern was the survival of local cultures and autonomy of communities (Esteva and Prakash, 1992). Amongst these, the SSP witnessed unprecedented and fierce opposition of a scale never witnessed in the history of water resources development. This makes it mandatory that any impact assessment should include the PAP and their changed conditions. Findings from such a study will have important lessons for future water projects in India and other developing economies, particularly those in Sub-Saharan Africa.

Chapter 2 reviewed some key international scientific literature dealing with the social, economic, and ecological impacts of large water resource systems. The review showed positive impacts of large water systems on regional economic growth, poverty reduction, social development, flood control, food security, and employment, though it noted that capital-intensive large water storage creation may not be the best option particularly in areas with surplus labor and small holdings. Further, benefit sharing would play a major role in the future of dams in developing countries, and success in realizing the potential poverty reduction impacts of large dams would depend on the nature of institutions being created to manage the dam-building projects. Further, the review showed that with changing technical environment, there is a need to redefine the objectives and criteria for assessing large water storages so as to reflect on the actual negative social and ecological challenges they pose. Also, there is a need to broaden the objective and criteria for evaluating the performance of large water systems, particularly in the context of developing economies, in order to capture the wide ranging positive indirect impacts, which often outweigh the direct ones.

Chapter 3 discussed the methodologies employed for estimating various impacts and benefits from the project, including the direct and indirect economic benefits of irrigation, domestic water supply, and power generation. It also presented the methodology for analyzing the economic value of the ecological benefit accrued from the release of Narmada water into the rivers. It also discussed the analytical procedure employed for quantifying the various physical and economic benefits, and sampling procedure and sample size for studying each type of impact, including the technical and economic externalities. Though the study employed was "longitudinal" in nature, involving pre- and post-Narmada scenarios, and used recall method to assess the wide ranging impacts of irrigation and drinking water supplies, the results vis-à-vis irrigation impacts were corroborated by the analysis of farming enterprise of those who have farms located outside the command area.

Chapter 4 discussed the key features of the Narmada River basin, the "award" made by the NWDT, the water and power systems being planned for the basin as per the master plan, water utilization plan for Gujarat, and the overall progress in the implementation of the SSP. The chapter also provided some analysis to show why a large water resource project like the SSP is extremely important for a state like Gujarat. The analysis shows that Gujarat has a very high regional variation in water resource endowment in terms of "per capita renewable water availability" from absolutely water-scarce north Gujarat region to water-abundant south Gujarat region. It also shows that water use is unsustainable in three out of the four regions. The water demand for 2025 far exceeds the total amount of utilizable water resources in the state and the magnitude of water scarcity would be serious in naturally water-scarce regions of the state as they have excessively high water demands.

Chapter 5 discussed the results of the analysis carried out using the data collected from several observation wells in the Narmada command area by SSNNL. The time series data corresponded to two different time periods, one before the introduction of Narmada water and the other post-Narmada. It showed that the introduction of surface water from Narmada canal for irrigation had resulted in changes in groundwater level trends over time, marked by relatively higher annual rise in water levels in the observation wells in the command area, except in Kheda district, during the post-Narmada period (2004–2010), as compared to the pre-Narmada period (1996–1999). Similarly, there has been a marked reduction in the salinity of groundwater over time, as indicated by relatively greater annual rate of decline in TDS during the post-Narmada period. These improvements in groundwater condition could be because of increased recharge from irrigation return flows and/or reduction in groundwater draft owing to the increased availability of canal water for irrigation. The findings suggest that it would be highly risky to generalize when it comes to analyzing the impact of canal irrigation on groundwater. Since water-level fluctuations are dynamic, it is important to carry out continuous monitoring of groundwater in the command area through a good network of observation wells.

Chapter 6 presented the results of the analysis on direct economic impact on farm households and wage laborers engaged in agriculture. The analysis presented in the chapter shows that the area under irrigation has increased substantially in all the selected locations after the introduction of water through Narmada canal. Maximum increase in the average irrigated area through gravity irrigation was found to be in Bharuch (from 1.27 ha to 5.11 ha). Impacts were sharper in the case of farmers irrigating through canal lift. In this respect, the maximum impact was found in Bharuch where the average area irrigated went up from 7.19 ha to 25.05 ha, followed by Ahmedabad where it went up from 8.23 ha to 14.17 ha.

Further, with the introduction of Narmada water, dependence on other sources of irrigation such as wells and purchased water has declined. In case of gravity irrigation, the highest reduction in irrigation by purchased water was found in Vadodara district. Similarly, well irrigation has become nonexistent in all the four selected locations except Ahmedabad. In the case of canal lift, the average area irrigated with purchased water went down drastically in all the selected locations, with zero area in all locations except Bharuch. In most of the study locations, the average gross cropped area per farmer increased after the introduction of water from Narmada canal. This shows the positive equity impacts canal irrigation can bring about in the region where access to irrigation wells is still skewed.

In the post-Narmada period, farmers shifted their cropping pattern toward remunerative crops. For instance, the gravity irrigators allocated greater proportion of their land to irrigated crops such as cotton (Bharuch), castor (Panchmahals and Vadodara), wheat, and maize in all study locations. The area under summer bajra too increased in three locations, viz., Narmada, Panchmahals, and Vadodara after the introduction of canal water. Similar changes in the cropping pattern were also noticed among the farmers doing irrigation through canal lift. Overall, there was a notable increase in the area under kharif paddy, cotton, castor, chick pea, wheat, and maize post the introduction of Narmada waters.

The study found remarkable increase in the crop yield (kg per ha) and net income returns per unit of land (₹ per ha) for both gravity and canal lift irrigators for all crops. The yield improvement was quite considerable for many crops, including castor, cotton, paddy, and wheat. This has been because of the following factors: (1) farmers providing irrigation to kharif crops, which were earlier grown under rain-fed conditions (such as chick pea, paddy, and cluster bean) with the number of irrigated crops increasing in all study locations; (2) farmers growing longer duration, high-yielding varieties of the crops, viz., castor and cotton, which require irrigation; (3) intensive use of labor and fertilizers along with irrigation, manifested by the increasing cost of cultivation. Thus, this is the composite effect of crop technology adoption, intensive use of labor, and intensive use of input, all of which were enabled by the provision of irrigation.

In both gravity and canal lift irrigated areas, the number of irrigated crops increased remarkably. The rain-fed crops were replaced by their irrigated counterparts. Interestingly, post the Narmada water, none of the locations had farmers raising crops under rain-fed conditions.

In spite of the increase in input costs for crop cultivation, there has been a remarkable increase in the net income from all crops across study locations for both canal irrigators and farmers doing irrigation using canal lift. The increase in the net income was found to be the highest for cotton, across the locations (gravity irrigation). In the case of lift irrigators, rise in the net income was less than that of gravity irrigators. This is mainly due to the higher increase in the input cost for canal lifting as it involves many costs, depending on the location of the farm vis-à-vis the canal, and whether the canal is in cutting or banking.

A marginal increase in the livestock holding was also observed in the command villages. The most significant rise was observed in Narmada district, where, on average, one in every two farming households has increased the holding by one animal. Similar trend was found in the case of farmers who are benefited by irrigation through canal lift in Ahmedabad, Mehsana, Narmada, Vadodara, and Bharuch. There were changes in the average annual income from dairy production in the villages located both in the canal command areas and areas irrigated through canal lifting.

Agricultural productivity growth also had a big impact on the employment scenario in the villages. There was an increased demand for farm labor especially in south Gujarat districts of Vadodara, Bharuch, and Narmada, and parts of Panchmahals due to the large expansion in the irrigated area owing to the availability of canal water. The increase in wage employment in agriculture post-Narmada times is significant for summer months, except in Bharuch.

The change in the net income per unit area of the crops, in the income from dairy farming per unit of livestock, the shift in cropping pattern, and the increase in the irrigated area had resulted in a major increase in the farm surplus of the farmers receiving Narmada waters. The increase in the overall farm surplus per ha of the gross cropped area in the gravity irrigated area was estimated to be varying from a lowest of ₹24,903 per ha per annum in Panchmahals to ₹48,348 per ha in Bharuch. In Vadodara, it was ₹48,126 per ha, and in Narmada, it was ₹41,475 per ha. This is a big gain, when we consider the fact that the areas which now receive water from Narmada canals by gravity mostly fall in the high rainfall areas.

As regards the farmers engaged in canal lifting, the increase in the overall farm surplus per ha of the gross cropped area was estimated to be varying from a lowest of ₹14,195 per ha in Narmada district to the highest of ₹109,855 per ha in Mehsana district. The exceptionally high income gain in the case of Mehsana is because of the high-valued crops such as fennel being grown by farmers in that area. But what is equally important is the fact that the gross cropped area of individual farmers is quite small here. The comparative analysis of gravity irrigators and canal lift irrigators shows that if greater control over water delivery is established, the farmers would be able to raise their income from farming, even if the cost of irrigation is high.

The increase in farm income had a direct impact on the household expenditure of the irrigators. There was a remarkable increase in the total household expenditure for the sample farm households which received Narmada canal water for irrigation. More importantly, the proportion of the expenditure for food items reduced considerably in all locations, a strong indication of reducing poverty. Similar trend was observed in the case of farmers lifting water form canals. The percentage expenditure on food has dropped in four out of the five locations post-Narmada waters. The households have also started spending more on children's education. The overall income of rural families, particularly farmers and wage laborers had also gone high post-Narmada.

Chapter 7 presented the findings with regard to the socioeconomic impacts of provision of water supplies from Narmada canal-based pipeline scheme in Saurashtra and Kutch. The study was carried out in both rural and urban areas of Jamnagar and Bhuj. The study found a marked change in people's reliance on the water supply sources post the Narmada drinking water supply. With the introduction of Narmada water, communities in rural areas of Jamnagar started using only Narmada source for drinking, cooking, domestic uses, and livestock drinking, whereas they used to depend on public bore wells and common stand posts prior to the introduction of Narmada waters. Similar trend was also observed in Jamnagar town, rural Bhuj, and Bhuj town.

But there was no significant improvement in the frequency of water supply post-Narmada, in spite of major improvement in water availability situation in terms of dependability. In both rural and urban areas of Bhuj, the frequency of water supply had actually reduced after the introduction of Narmada canal-based water supply. However, the overall physical access to water sources had improved after the introduction of Narmada waters. The survey showed that greater proportion of the sample households get water within the dwelling or within the premises after the introduction of Narmada piped water supply scheme.

Further, after the introduction of Narmada water, the households could save time on water collection. Time saving was slightly higher during the summer season, particularly in the case of rural Bhuj and Jamnagar town, where the magnitude and extent of water shortage used to increase during summer prior to the Narmada piped water. Overall, the reduction in time saving was only marginal because of the larger number of trips that family members had to undertake to collect water as a result of reduction in the frequency of water supply. The communities were able to productively use the time saved from water collection.

While there was no significant increase in water use at the household level for domestic and livestock purposes in volumetric terms during post-Narmada, there was a marked improvement in the quality of water supplied. But what is equally important is the fact that adequate quantity of water is being supplied from the single source of the Narmada scheme. As research at the global level has shown, such quantities of household water use are capable of bringing about positive health benefits, if water is of good quality. The study also points to the fact that in the absence of a dependable source of safe water, households often manage some of their household water needs such as washing clothes, cleaning utensils, and sanitation with inferior quality water.

Water quality improvement resulted in perceptible reduction in average annual expenditure on health, lower incidence of water-related health problems, and reduction in the number of days of employment lost due to water-related sickness. However, some long term studies based on "water supply surveillance" will be required to establish the impact of the provision of safe water supplies on socioeconomic and health status of households, particularly those on child health, nutrition, education, and incidence of water-borne diseases.

Chapter 8 discussed the indirect impacts of canal irrigation. It quantified the economic value of the benefits from the following positive externalities of canal irrigation: reduction in energy use in well irrigation in the command area and incidence of well failures owing to the rise in water table; increase in area irrigated by wells and yield and income from well-irrigated crops owing to the improvement in yields of agro wells; improved sustainability of drinking water supply sources due to improvement in their yield; and the increase in wage rates in rural areas owing to increased demand for wage labor in agriculture. It employed some unique methodologies in environmental economics to quantify the indirect benefits.

First of all, based on micro-level field data, it showed that irrigation wells located in the command area of Narmada canal were benefited by seepage from canal water. This is manifested by a consistent rise in groundwater levels across seasons, resulting in the reduced depth of pumping. Therefore, the results corroborate with the findings of the macro-level analysis done using secondary data. Consequently, there has been a notable reduction in the incidence of well failures in the canal command area and the increase in the command areas of wells post-Narmada. But the same trend was not visible in Panchmahals district. This was probably due to a substantial increase in the number of wells tapping the hard rock aquifers of the district over the past few years, which perhaps is disproportionately higher than the additional recharge available from the small canal irrigated area.

After the introduction of Narmada waters, the cropping pattern of well irrigators in the canal command had vastly changed. There was a substantial increase in the yield and income returns per unit of land (₹ per ha). While the area under the "normally" rain-fed crops of kharif season reduced, the area under irrigated crops such as cotton, castor, and wheat had increased in some locations. Almost every crop grown in the well commands showed yield changes, mostly positive after the introduction of Narmada waters. The increase in the yield and income from crops in the well commands could be explained by the greater dosage of irrigation to these crops, which is reflected in the increase in the average volume of groundwater applied by the farmers per ha of land post-Narmada.

The indirect benefit from canal irrigation in the form of the reduced economic cost of energy used for pumping groundwater for irrigation was estimated to be huge. The benefit was lowest in Bharuch and highest in Mehsana, with variation being wide. Such huge variations occur because of three factors: (1) major variations in the rise in the groundwater table observed across regions, which cause proportional variation in the energy saving for pumping a unit volume of groundwater and (2) major variation in the amount of groundwater used per ha of the well-irrigated area, owing to variations in climate. These foregoing analyses suggest that analyzing the performance of gravity irrigation systems based on the area irrigated by the canals will be highly misleading. Return flows from irrigated fields and canal seepage can enhance the overall economic benefits from surface irrigation systems, by promoting well irrigation and by reducing the energy cost of pumping groundwater.

Another major indirect impact of canal irrigation is the rise in wage rates in rural areas. The increased demand for farm labor had resulted in a substantial increase in wage rates in the selected locations post the Narmada water introduction, which was found to be disproportionately higher than the historical increase in wage rates in other areas of Gujarat that are not so much benefited by Narmada canals.

The indirect economic impact of canal irrigation in the form of improved sustainability of groundwater-based drinking water supply schemes in south and east Gujarat parts of Narmada command for a population of five million was estimated to be ₹1,032.5 million per annum, whereas the indirect economic impact of Narmada canal-based piped water supply to a total population of 24.33 million in Saurashtra, Kutch, and north and central Gujarat owing to the saving in energy used for pumping groundwater for rural and urban water supply was estimated to be ₹857.7 million per annum.

Chapter 9 discussed the environmental externalities of the SSP. It mainly covered three aspects: ecological gains from the release of water into the environmentally stressed rivers, the economic value of the environmental benefits due to reduced carbon emission from energy production, and the environmental impacts of canals such as waterlogging and salinity. The indirect economic impact of clean energy production for an energy-scarce state through hydropower from the SSP was estimated to be ₹1.61 billion per annum for a total estimated average energy production of 3,436 million units per annum. This is equal to the cost incurred for mitigating the carbon emission that is likely to be caused by the use of fossil fuel (coal based) for producing equivalent amount of electricity from thermal power stations, which is the only alternative energy production system available for the region.

While waterlogging was one of the major concerns raised by environmentalists and development activists alike at the project planning stage, there was no incidence of waterlogging reported in the command area of the SSP, though water levels have risen in most parts of the command.

This is because most parts of the command area had experienced groundwater overdraft for 2–3 decades. Further, the future possibility of waterlogging appears to be less due to the following reasons: (1) the average depth of watering from the SSP is likely to come down in the coming years, as the second phase of the command area becomes ready; (2) once the dosage of surface water reduces, the groundwater withdrawal in the command area would increase both in the south and north, as canal water alone cannot meet the water requirement of crops grown in two seasons, further reducing the possibility of water table build up; and (3) as the SSP will not be able to fully meet the demand for irrigation water in the regions where canal network is extended, the deficit has to be met from groundwater, and the use of the same would increase over time, with population growth.

Chapter 10 discussed the changes in the lives of people who were displaced from the valley due to reservoir submergence. The chapter provided a quick narrative of the struggle in the valley for securing a just rehabilitation policy from the Government of Gujarat under the SSP, and the policy and norms finally adopted by the government in 1987. The impact of rehabilitation on the socioeconomic dynamic of the resettled families was analyzed in different time periods extending over two decades. For that the study reviewed the surveys carried out in 1994 (Sah, 1994), the recent literature on the theme (Shah and Tomar, 2011), and the field work carried out in a few settlements by the authors. Our review found that significant proportion of the displaced families(that is, 10,065 out of the 11,029 families) were rehabilitated in Gujarat, and these families would finally settle down in 236 new settlements in south and central Gujarat.

The displaced families, which have got 5 acres of land each, have moved away from subsistence farming to market-oriented cultivation. The livelihoods of the affected families have now become heavily dependent on agriculture. The overall agricultural income of the affected families have increased after relocation to the new sites owing to the increased yield of the crops grown and the shift toward commercial crops. But they still earn less than the families in the host villages in spite of having larger operational holdings, as their adoption of improved crop and input use technologies was much lower. In lieu of the adoption of crops that require higher dosage of inputs including water, the income from farming has become more vulnerable to droughts.

Chapter 11 discussed the alternatives to enhance the social, economic, and ecological benefits from the SSP. The authors argued that there are better ways of maximizing social, economic, and ecological benefits from the SSP, which called for a paradigm shift in the way irrigation efficiencies in canal systems are assessed. The chapter argued that the uncommitted and unutilized monsoon flow into the Sardar Sarovar reservoir should be transferred to north and central Gujarat region, and the tertiary canals in the alluvial areas of north and central Gujarat should be left unlined to maximize the return flows from seepage and irrigated fields as it would result in larger social benefits. At the same time, in the water-rich shallow groundwater areas of south Gujarat, the canals should be lined to conserve water and to prevent problems of waterlogging. The authors have argued that defining water entitlements in volumetric terms for the farmers would encourage title holders, especially those in south Gujarat to use water very efficiently for crop production, and to engage in trading with the saved water. Large-scale interregional water trading should be promoted in the command area through the farmer organizations created at different levels of the hydraulic system hierarchy.

Will the Positive Impacts Be Sustainable?

At present, the SSP is using around 3.41 MAF of water to irrigate nearly 0.6 million ha of land through direct gravity irrigation and canal lift. This leaves a water application dosage of nearly 55 cm (22 inches) of water. With this water, in conjunction with the use of some water from wells in the command area, the farmers are able to take irrigated winter crops, along with longer duration varieties of kharif crops with the help of supplementary irrigation. Once, the canal network for the entire designated command is ready, the average delta for the command area would reduce to around 18 inches, after factoring in the system losses. Nevertheless, this is not likely to impact on the current farming enterprise because of two reasons. First, surplus water from upper basin areas would be available in excess of the 9.75 MAF of water already allocated to Gujarat, until all the major and medium reservoirs in MP part of the basin are completed. This would flow into the terminal reservoir of Sardar Sarovar in the Narmada basin and could be diverted through the SSP canal network to the command area farmers for kharif irrigation. Second, there has been substantial "buildup" of groundwater in the canal command area and area irrigated through canal lift, owing to the return flows from surface irrigation. The additional groundwater will be available for use by the farmers in the designated command area in conjunction with the surface water allocated through canals. Hence, they would be able to sustain irrigated crop production with at least two crops.

Conclusions and Policy

There were no studies in the past which comprehensively analyzed the direct and indirect impacts of a large, multipurpose water resource system like the SSP, in quantitative terms. This is the first one of its kind analyzing the positive externalities as well as the direct social, economic, socioeconomic, and ecological impacts of a multipurpose water resource project that too when the project is still under implementation. The direct economic impacts of canal irrigation are well documented. All those benefits, in terms of increased income from crop and milk production are quite impressive in the SSP command, including the area irrigated by canal lifting. They also resulted in better living conditions for the farm households, manifested by increased household expenditure on food. One important socioeconomic fallout in this context is the increased demand for wage labor in agriculture and greater employment for laborers in rural areas. The poverty-reduction impacts of canal irrigation, on the farmers and farm laborers, inside and outside the designated command, are huge.

Contrary to the common belief, such big gains have not come from the increase in cropping intensity, except in places which have not only rain-fed agriculture because of saline groundwater, but rain-fed crops replaced by irrigated counterparts. In all the surveyed locations, farmers were found to be irrigating almost every crop, be it kharif or winter; they also raise long-duration and high-yielding irrigated varieties for crops like castor and cotton, replacing the short duration rain-fed ones, and provide more agronomic inputs, thereby substantially raising the yield and net income. In the saline areas, where no sources of irrigation were available prior to Narmada, a significant increase in the cropping intensity was seen, with farmers taking crops in 2–3 seasons. In such cases, the economic impact on farm households was remarkable.

The indirect economic impacts of irrigation are too large to be ignored. These impacts get affected through reduced energy use for well irrigators owing to the rise in groundwater levels; improved sustainability of drinking water sources based on groundwater supplies owing to yield improvements and quality amelioration, benefiting an estimated 5 million people; reduced incidence of well failures in command areas; boost in income of well irrigators in command area from crops and livestock through yield increase resulting from intensive use of groundwater for irrigation and better availability of biomass as fodder; and remarkable increase in wage rates of laborers in rural areas across seasons, owing to increased demand for agricultural labor.

The drinking water supply scheme based on Narmada canal has played a significant role in improving the access of the village and urban communities to reliable and dependable sources of good quality water supplies. While the distance travelled by women to fetch water reduced significantly in most instances, there has been some reduction in time spent in water collection as well. While people used to depend on multiple sources of water, prior to Narmada even compromising on the quality, with the introduction of piped water supply, all the water needs are met from a single source. Though there has not been any notable difference in the quantity of water used by the communities for domestic purpose, the quantum of water used in per capita terms remains adequate to maintain a healthy living. The amount of safe water used by the households had substantially increased after the Narmada canal-based pipeline scheme. On the social front, the improved availability of good quality water for consumption had impacted on the health status as indicated by the reduction in the incidence of water-borne diseases.

The indirect economic impacts are: reduced energy cost of production and supply of water for domestic and municipal uses, with the replacement of groundwater-based sources by Narmada canal-based pipeline in rural and urban areas, estimated to be around ₹857.7 million per annum for an population of about 24 million people, and the ability of men and women to reach their work site on time, and increase in their employment opportunities. Lastly, the indirect economic benefits of producing clean energy through hydropower, which can defer the investments in fossil fuel-based power generation, thereby saving the cost of capturing emitted carbon is worked out to be more than ₹1.6 billion annually.

All these benefits and impacts are realized with just one-third of the designated command areas receiving Narmada water and a lesser percentage of the planned villages receiving drinking water supplies from the Narmada canal-based drinking water scheme. Some of the direct and indirect economic benefits are likely to magnify in proportion to the increase in irrigated area from canals and increase in the coverage of drinking water supply schemes, some of the other indirect benefits are likely to magnify with irrigated area expansion. These indirect benefits contribute to the multiplier effects of irrigation. The multiplier effects of irrigation that are likely to magnify over a period of time would essentially come from growth in agro-processing industries such as rice and flour mills and cold storages. This would ultimately have a huge poverty-reduction impact across Gujarat.

Learning for Other Developing Countries

There are many developing economies around the world, whose water resources are underutilized. Many countries in Sub-Saharan Africa fall in this category (Kumar et al., 2013; World Bank, 2004, 2006a, 2006b). The per capita reservoir storage in these countries is very poor, when compared to the developed countries, as shown in Figure 12.1 (Kumar, 2009).

Development of reservoir storage has an important role in improving the access to and use of water; the two prerequisites for improving the water situation of a region, though intensive water development in river basins might cause environmental water stress reducing the values of water environmental index (Shah and Kumar, 2008; Kumar, 2009). This is evident from the direct power relationship between storage development and water security (expressed in terms of SWUI) (Figure 12.2).

Figure 12.1:
Per Capita Annual Reservoir Storage (m³) in Selected Countries

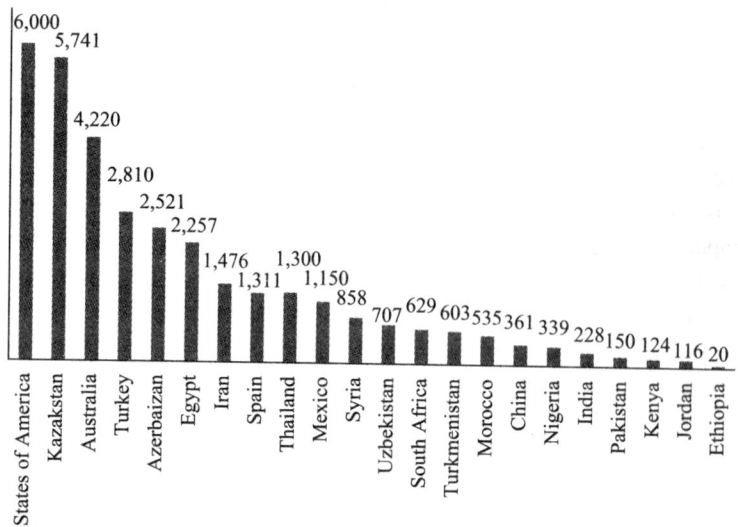

Source: Kumar, 2009.

Figure 12.2:
Sustainable Water Use Index versus per Capita Reservoir Storage

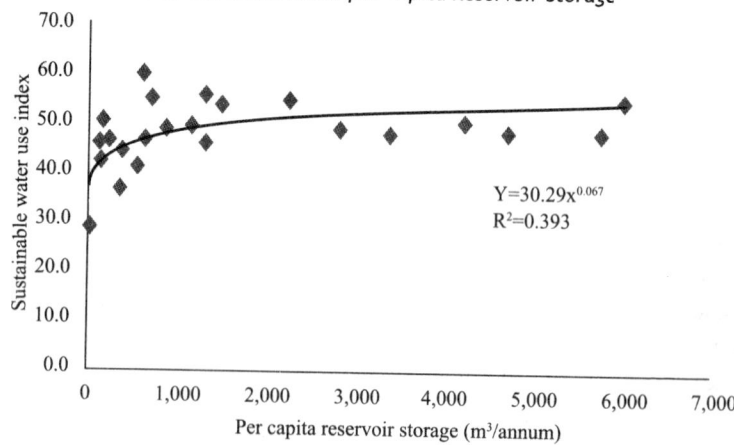

Source: Kumar, 2009.

The R^2 value was estimated to be 0.39. Figure 12.2 shows that countries having higher per capita reservoir storage (expressed in m^3 per capita per annum) have higher values of SWUI. Major improvements in water security occurred within the range of 0–1,500 m^3 per capita per annum, and leveling off thereafter.

However, the amount of storage that needs to be created to improve access to and use of water depends on the type of climatic conditions. It is also important to note that access to arable land would also be an important factor determining the storage requirements, as water requirement for agriculture would change with this. In temperate and cold climates, the demand of irrigation, the largest water use sector, would be considerably small as compared to tropical and hot climates. Hence, the storage requirements in such regions would be much lower and would be mainly limited to that for meeting domestic/municipal water needs and water for manufacturing. Therefore, it is logical to explore links between storage development for meeting various human needs and economic growth only in tropical and hot climates.

But the sheer scale of water infrastructure in rich countries is not widely recognized and appreciated (HDR, 2006, p. 155). Many developed countries of the world that experience tropical climates had high water storage in per capita terms. The United States had a per capita storage capacity of nearly 6,000 m^3. Australia had a total storage of 79,000 MCM per annum create through 447 large dams, providing per capita water storage of nearly 3,808 m^3 per annum. Aquifers supply another 4,000 MCM per annum. China has a per capita reservoir storage capacity of 2,000 m^3 per annum through dams, and an actual storage of nearly 360 m^3 per capita (Kumar, 2009).

Like in India, governments in many countries of Africa and South and South East Asia are facing intense resistance from local communities and environmental groups against large water projects, for their potential negative social and environmental impacts (WCD, 2000). This is dampening the much needed investments in water infrastructure projects in these countries. This is occurring at a huge social (opportunity) cost (Priscoli and Briscoe, 2011). They are many intriguing reasons for this growing opposition to large water projects. One of them is geopolitical. As noted by John Briscoe of the World Bank in an interview with the Editor-in-Chief of *Water Policy*, "The fashion of development lending had shifted to social issues as embodied in the Millennium Development Goals which prioritize health, education, gender and water and sanitation services" (Priscoli and Briscoe, 2011: p. 148). In the process, water infrastructure, as a precondition for economic growth, had essentially been relegated through a process led by the rich countries which already have the infrastructure (Priscoli and Briscoe, 2011).

Another important reason is the lack of capacity on the part of the governments in these countries and their donors to articulate all the social, economic, and environmental benefits from such projects, particularly the indirect and intangible ones and quantify them. The result is that the project benefits do not appear to sufficiently outweigh the costs. The beneficiaries also appear to be skewed. Failure to adopt rigorous and convincing methodologies for benefit–cost evaluation and the lack of the ability to foresee the likely positive changes on the physical and socioeconomic systems constitute some of the associated problems. The third reason is the lack of provision in the project design for paying adequate compensation to the affected communities for resettlement and rehabilitation, which again is linked to their inability to foresee the benefits (Shah and Kumar, 2008).

The advantage in improved sustainability of well irrigation is particularly important for Sub-Saharan Africa. The drought-prone areas of western, eastern, and southeastern Africa have two distinct geological formations: (1) shallow, patchy, minor aquifer system of low storage, with weathered crystalline basement and (2) consolidated sedimentary rocks which form generally deeper, but less extensive and geologically more complex, aquifers (Foster *et al.*, 2006). There is paucity of reliable information on the availability and use of groundwater in these Sub-Saharan African countries (Foster *et al.*, 2006). The patchy information that is available from various sources suggests that currently only a small fraction of the internal groundwater resources in Sub-Saharan Africa is used. The consumptive use of groundwater for irrigation is only 0.2 percent of the internal groundwater, while it is as high as 106 percent for northern Africa (Siebert *et al.*, 2010).

Analysis shows that the per capita consumptive groundwater in the regions of Sub-Saharan Africa is one of the lowest in the world. While in South Asia, which is known for intensive groundwater use for irrigation, it is 164 m^3 per capita, it is 80.2 m^3 per capita in northern Africa.

It varies from a miniscule of 0.38 m³ per capita in east Africa to 13.4 m³ per capita in southern African countries.[1] The constraints for the low level of exploitation of groundwater in Sub-Saharan Africa are the local level uncertainty about the resource occurrence, high drilling costs, and shortage of capital. Gravity irrigation for long time in a particular locality could help remove the prevailing uncertainty about the availability of groundwater, as it would improve groundwater recharge and raise water levels. This would encourage greater private investments in well irrigation.

The fact that in quantitative terms these benefits can be far higher than those considered in normal benefit–cost calculations should encourage the official agencies to build all the costs in the project design, and adequately compensate the affected communities by recovering the same from all the beneficiaries. Only such a paradigm shift in project planning can eliminate the current impasse. In many instances, as seen in the case of the SSP, the energy sector, the drinking water supply sector, and groundwater irrigation sector are large beneficiaries of surface irrigation. Identifying all the beneficiaries and quantifying the benefits would, however, require sophisticated methodologies, which would offer new sets of challenges.

[1] Authors' own analysis based on Siebert *et al.*, 2010, and population figures for these regions.

Bibliography

Aiga, H. and T. Umenai (2002) "Impact of Improvement of Water Supply on Household Economy in a Squatter Area of Manila," *Social Science & Medicine*, 55: 627–641. Available at http://dx.doi.org/10.1016/S0277-9536(01)00192-7.

Alagh, Y. K. (2010) "A Sardar Sarovar Riddle," *Financial Express*, April 19.

Alagh, Y. K., R. D. Desai, G. S. Guha, and S. P. Kashyap (1995) *Economic Dimensions of Sardar Sarovar Project*, New Delhi: Har-Anand Publications.

Allen, R. G., L. S. Willardson, and H. Frederiksen (September 11–12, 1998) "Water Use Definitions and Their Use for Assessing the Impacts of Water Conservation." Proceedings of ICID Workshop on Sustainable Irrigation in Areas of Water Scarcity and Drought, edited by J. M. de Jager, L. P. Vermes, R. Rageb, Oxford University Press, England, pp. 72–82.

Amarasinghe, Upali, B. R. Sharma, N. Aloysius, C. Scott, V. Smakhtin, and C. de Fraiture (2004) "Spatial Variation in Water Supply and Demand Across River Basins of India," Research Report No. 83, International Water Management Institute, Colombo, Sri Lanka.

Arghyam/IRAP (2010) "Integrated Urban Water Management for Different Urban Typologies of India," Final report submitted to Arghyam, Bangalore.

Baviskar, A. and A. K. Singh (1994) "Malignant Growth: The Sardar Sarovar Dam and Its Impact on Public Health," *Environmental Impact Assessment Review*, 14(5–6): 349–358.

Berger, Thomas R. (1994) "The Independent Review of the Sardar Sarovar Projects 1991–1992," *Impact Assessment*, 12(1): 3–20.

Bhattarai, M., R. Sakthivadivel, and I. Hussain (2002) "Irrigation Impacts on Income Inequality and Poverty Alleviation: Policy Issues and Options for Improved Management of Irrigation Systems," Working Paper No. 39, International Water Management Institute, Colombo, Sri Lanka.

Biswas, A. K. and Cecelia Tortajada (2001) "Development and Large Dams: A Global Perspective," *International Journal of Water Resources Development*, 17(1): 9–21.

Biswas, A. K., O. Unver, and C. Tortajada (2004) *Water as a Focus for Regional Development*, New Delhi: Oxford University Press.

Brewer, J., Shashi Kolavalli, A. H. Kalro, G. Naik, S. Ramnarayan, K. V. Raju, and R. Shakthivadivel (1999) *Irrigation Management Transfer in India: Policies, Processes and Performance*, New Delhi/Calcutta: Oxford University Press/IBH.

Briscoe, John (February 26–29, 2005) "India's Water Economy: Bracing up for a Turbulent Future," Key note address at the 4th Annual Partners' Meet of IWMI-Tata Water Policy Research Program, Institute of Rural Management, Anand.

——— (2011) "Invited Opinion Interview: Two Decades at the Center for World Water Policy," *Water Policy*, 13(2): 147–160.

Business Line (2009) "Gujarat Emerges a Silver Lining," August 8.

Cairncross, S. and J. Cliff (1986) "Water Use and Health in Mueda, Mozambique," *Transactions of the Royal Society of Tropical Medicine and Hygiene*, 98: 419–427.

Cairncross, S. and R. Feachem (1993) *Environmental Health and Engineering in the Tropics*, 2nd edition, UK: John Wiley & Sons.

Chakravorty, Ujjayant and Chieko Umetsu (2003) "Basin Wide Water Management: A Spatial Model," *Journal of Environmental Economics and Management*, 45(2003): 1–23.

Chambers, R. (1988) *Managing Canal Irrigation: Practical Analysis from South Asia*, New Delhi: Oxford University Press.

Chetwynd, Eric, Daniel Dworkin, and Son Ung Kim (1981) "Korean Potable Water System Project: Lessons from Experience," Project Impact Evaluation Report No. 20, US Agency for International Development.

Chowdhury, Anis (2011) "Food Price Hikes: How Much Is Due to Excessive Speculation?" *Economic and Political Weekly*, 46(28): 12–15.

D'Souza, D. (2002) *Narmada Dammed: An Enquiry into the Politics of Development*, New Delhi: Penguin Books India.

Daines, S. R. and J. R. Pawar (1987) "Economic Returns to Irrigation in India, SDR Research Groups Inc. & Development Group Inc." Report prepared for the U.S. Agency for International Development Mission to India, New Delhi, India.

Das, Keshab and Ruchi Gupta (2010) "Management by Participation?: Village Institutions and Drinking Water Supply in Gujarat," Working Paper 158, Gujarat Institute of Development Research (GIDR), Ahmedabad, Gujarat.

De Fraiture, Charlotte and Chris Perry (June 15–17, 2002) "Why is Irrigation Water Demand Inelastic at Low Price Ranges?" Paper presented at the Conference on Irrigation Water Policies: Micro and Macro Considerations, Agadir, Morocco.

Desai, S. J. and M. B. Joshi (2008) *Narmada Water Plays Its Role-Capturing Initial Trends from Gujarat*, Gujarat: Sardar Sarovar Narmada Nigam Limited.

Dharmadhikary, S. (1993) "Hydropower from Sardar Sarovar: Need, Justification and Viability," *Economic and Political Weekly*, 28(48): 2584–2588.

———— (2005) *Unraveling Bhakra: Assessing the Temple of Resurgent India*, New Delhi: Manthan Adhyayan Kendra.

Dhawan, B. D. (ed.) (1990) *Big Dams: Claims, Counter Claims*, New Delhi: Commonwealth Publishers.

Dholakia, Ravindra H. (2011) "Review of Social Cost–Benefit Analyses for the Sardar Sarovar Project," in Volume 1 of *Sardar Sarovar Project on the River Narmada: History of Design, Planning and Appraisal*, edited by R. Parthasarathy and Ravindra H. Dholakia, Ahmedabad: CEPT University Press.

Dholakia, Ravindra and Samar Dutta (2010) *High Growth Trajectory and Structural Changes in Gujarat*, edited by Ravindra Dholakia and S. Dutta, New Delhi: MacMillan Press.

Drexhage, John and Deborah Murphy (2010, September), "Sustainable Development: From Brundtland to Rio 2012." Available at http://www.un.org/wcm/webdav/site/climatechange/shared/gsp/docs/GSP1-6_Background%20on%20Sustainable%20Devt.pdf (December 22, 2011).

Esrey, S. A., R. G. Feachem, and J. M. Hughes (1985) "Interventions for the Control of Diarrheal Diseases Among Young Children: Improving Water Supplies and Excreta Disposal Facilities," *Bulletin of the World Health Organization*, 63(4): 757–772.

Esrey, S. A., J. B. Potash, L. Roberts, and C. Shiff (1991) "Effects of Improved Water Supply and Sanitation on Ascariasis, Diarrhea, Dracunculiasis, Hookworm Infection, Schistosomiasis, and Trachoma," *Bulletin of the World Health Organization*, 69(5): 609–621.

Esteva, G. and M. Prakash (1992) "Grassroots Resistance and Sustainable Development," *The Ecologist,* 22(2): 45–51.

Esther, Duflo and Rohini Pande (2007 [2005]) "Dams," *The Quarterly Journal of Economics* (MIT Press), 122(2): 601–646.

Evenson, R., C. Pray, and M. Rosegrant (1999) "Agricultural Research and Productivity Growth in India," Research Report No. 109, International Food Policy Research Institute, Washington, DC, USA.

FAO (2003) "Preliminary Review of the Impact of Irrigation and Poverty with Special Emphasis on Asia," Report No. AGL/MISC/34/2003, Rome.

Feachem, R., et al. (1997) "Lesotho Village Water Supplies: An Ex-post Evaluation," Final Report to the Ministry of the Overseas Development Administration, Great Britain.

Fisher, William. F. (2001) "Diverting Water: Revisiting Sardar Sarovar Project," *Water Resources Development*, 17(1): 303–314.

Foster, S., A. Tuinhof, and H. Garduño (2006) "Groundwater Development in Sub-Saharan Africa: A Strategic Overview of Key Issues and Major Needs," Case Profile Collection No. 15, World Bank, Washington, DC, USA.

Frederick, K. D. (1992) "Balancing Water Demand with Supplies: The Role of Management in a World of Increasing Scarcity," Technical Paper No. 189, World Bank, Washington, DC, USA.

Gleick, Peter H. (2000) "The Changing Water Paradigm: A Look at Twenty-First Century Water Resources Development," *Water International*, 25(1): 127–138.

Government of Gujarat (GOG) (1989) *Planning for Prosperity: Sardar Sarovar Development Plan*, Gandhinagar: Sardar Sarovar Narmada Nigam Limited.

——— (1991) *All About Narmada: Directorate of Information*, Gandhinagar: Government of Gujarat.

——— (1996) *Water Resources Planning for the State of Gujarat*, Gandhinagar: Narmada and Water Resources Department.

Government of India (GOI) (1979) "Final Order and Decision of the tribunal." Available at http://www.sardarsarovardam.org/assets/SitepagesDocument/SPD__Userid2_20100402_104848.pdf (accessed December 20, 2011).

——— (1999) *Report of the National Planning Commission: Integrated Water Resources Development*, Government of India.

Government of Madhya Pradesh (2001) *Irrigation Statistics for Year 2000–2001*, unpublished report, Government of Madhya Pradesh, Bhopal.

Goyal, S. P., B. Sinha, N. Shah, and H. S. Panwar (1999) "Sardar Sarovar Project: A Conservation Threat to the Indian Wild Ass *(Equus hemionus khur),*" *Biological Conservation*, 88(2): 277–284.

Grey, D. and C. Sadoff (2005) "Cooperation on International Rivers: A Continuum for Securing and Sharing Benefits," *Water International*, 30(4): 420–427.

——— (2007) "Sink or Swim: Water Security for Growth and Development," *Water Policy*, 9(6): 541–570.

Gujarat Ecology Commission (GEC) (1997) *Eco Regions of Gujarat*, Vadodara: Gujarat Ecology Commission.

Gujarat State Drinking Water Infrastructure Co Ltd (2000). "Gujarat Jal-Disha 2010: A Vision of a Healthy and Equitable Future with Drinking Water, Hygiene and Sanitation for All," A report by a working group constituted by Gujarat State Drinking Water Infrastructure Co. Ltd.

Hernandez-Mora, Nuria, R. Llamas, L. Martinez-Cortina (1999) "Misconceptions in Aquifer Overexploitation Implications for Water Policy in Southern Europe." Paper for the 3rd Workshop SAGA, Milan.

Hirway, Indira and Neha Shah (2011) "Labour and Employment Under Globalisation: The Case of Gujarat, Review of Labour," *Economic and Political Weekly*, XLVI(22): 57–65.

Howard, Guy and Jamie Bartram (2003) *Domestic Water Quantity, Service Level and Health*, Geneva: World Health Organization.

Howe, Charles W., Dennis R. Schurmeier, and W. Douglas Shaw, Jr., (April, 1986) "Innovative Approaches to Water Allocation: The Potential for Water Markets," *Water Resources Research*, 22(4): 439–445.

Howell, Terry (2001) "Enhancing Water Use Efficiency in Irrigated Agriculture," *Agronomy Journal*, 93(2): 281–289.

Human Development Report (HDR) (2006) *Human Development Report-2006*, New York: United Nations.

Hussain, Intizar and Munir Hanjra (2003) "Does Irrigation Water Matter for Rural Poverty Alleviation? Evidence from South and South East Asia," *Water Policy*, 5(5): 429–442.

Institute of Rural Management Anand (IRMA)/UNICEF (2001) "White Paper on Water in Gujarat," Report submitted to Government of Gujarat, Gandhinagar.

Irajpoor, A. A. and M. Latif (2011) "Performance of Irrigation Projects and Their Impacts on Poverty Reduction and Its Empowerment in Arid Environment," *International Journal of Environmental Science and Technology*, 8(31): 533–544.

Iyer, R. Ramaswamy (2012) "Bridge over the River Cauvery." Lead Opinion in *The Hindu*, October 16.

Jacob, Granit and L. Andreas (December, 2009) "The Role of Large Scale Artificial Water Storage in the Water—Food—Energy Development Nexus," Summary of the policy report for SIDA.

Jain, L. C. and S. C. Behar (2008) *Performance and Development Effectiveness of the Sardar Sarovar Project*, pp. 31–32, Mumbai: Tata Institute of Social Sciences.

Jha, Nitish (2010) "Access of the Poor to Water Supply and Sanitation in India, Salient Concepts, Issues and Cases," Working Paper No. 62, International Policy Centre for Inclusive Growth (IPC-IG), Brazil.

Joshi, Mukesh B. and Vivek Kapadia (2010) "Sharing Water in the 21st Century: Rethinking the Rationale," *Irrigation and Drainage*, 59 (1): 92–101.

Kavalanekar, N. B., S. C. Sharma, and K. R. Rushton (1992) "Over-exploitation of an Alluvial Aquifer in Gujarat, India," *Hydrological Sciences Journal*, 37(4): 329–346.

Kay, Melvyn, T. Franks, and L. Smith (1997) *Water: Economics, Management and Demand*, London: E&FN Spon.

Koppen, B. van, S. Smits, P. Moriarty, F. Penning de Vries, M. Mikhail, and E. Boelee (2009). *Climbing the Water Ladder: Multiple-use Water Services for Poverty Reduction*, TP series, No. 52, p. 213. The Hague, the Netherlands: IRC International Water and Sanitation Centre and International Water Management Institute.

Kothari, A. and Rajiv Bhartari (1984) "Narmada Valley Project: Development or Destruction?" *Economic and Political Weekly*, 19(22–23): 907–909.

Kumar, M. Dinesh (2002) "Reconciling Water Use and Environment: Water Resources Management in Gujarat Resource, Problems, Issues, Options, Strategies and Framework for Action," Report of the Hydrological Regime Subcomponent of the State Environmental Action Programme supported by the World Bank, Gujarat Ecology Commission, Vadodara.

————— (2003). "Food Security and Sustainable Agriculture in India: The Water Management Challenge." International Water Management Institute (IWMI) Working Paper 60, pp. 32–36.

————— (2005) "Impact of Electricity Prices and Volumetric Water Allocation on Energy and Groundwater Demand Management: Analysis from Western India," *Energy Policy,* 33(1): 39–51.

————— (2007) *Groundwater Management in India: Physical, Institutional and Policy Alternatives*, New Delhi: SAGE Publications.

————— (2009) *Water Management in India: What Works, What Doesn't*, New Delhi: Gyan Books.

————— (2010) *Managing Water in River Basins: Hydrology, Economics, and Institutions*, New Delhi: Oxford University Press.

Kumar, M. Dinesh and Upali Amarasinghe (eds) (2009) *Water Productivity Improvements in Indian Agriculture: Potentials, Constraints and Prospects, Strategic Analysis of National River Linking Project of India Series 4*, Colombo: International Water Management Institute.

Kumar, M. Dinesh, Vishwa Ballabh, and Jayesh Talati (2000) "Augmenting or Dividing: Surface Water Management in the Water-scarce River Basin of Sabarmati," IRMA Working Paper 147, Institute of Rural Management, Anand.

Kumar, M. Dinesh and N. Bassi (2011) "Maximizing the Social and Economic Returns from Sardar Sarovar Project: Thinking beyond the Convention," in *Sardar Sarovar Project on the River Narmada: Impacts So Far and Ways Forward,* edited by R. Parthasarathy and R. Dholakiya, pp. 747–776, Delhi: Concept Publishing Company Pvt. Ltd.

Kumar, M. Dinesh, Ajaya Kumar Malla, and Sushanta Tripathy (2008) "Economic Value of Water in Agriculture: Comparative Analysis of a Water-Scarce and a Water-Rich Region in India," *Water International*, 33(2): 214–230.

Kumar, M. Dinesh, A. Narayanamoorthy, and M. V. K. Sivamohan (2010a) "Pampered Views and Parrot Talks: In the Cause of Well Irrigation in India," Occasional Paper No. 1, Institute for Resource Analysis and Policy, Hyderabad.

Kumar, M. Dinesh, A. Narayanamoorthy, O. P. Singh, M. V. K. Sivamohan, Manoj Sharma, and Nitin Bassi (2010b) "Gujarat's Agricultural Growth Story: Exploding Some Myths," Occasional Paper No. 2, Institute for Resource Analysis and Policy, Hyderabad.

Kumar, M. Dinesh and Rahul Ranade (July, 2004) "Large Water Projects in the Face of Hydro Ecological and Socio-economic Changes in Narmada Valley: Future Prospects and Challenges," *Journal of Indian Water Resources Society*, 39(3).

Kumar, M. Dinesh and R. Ranade (2004) "Narmada Water for Groundwater Recharge in North Gujarat: Conjunctive Management in Large Irrigation Projects," *Economic and Political Weekly*, 39(31): 3510–3513.

Kumar, M. Dinesh, Z. Shah, S. Mukherjee, and A. Mudgerikar (2008) "Water, Human Development And Economic Growth: Some International Perspectives," in *Managing Water in the Face of Growing Scarcity, Inequity and Declining Returns: Exploring Fresh Approaches*, edited by M. D. Kumar, pp. 842–858, Hyderabad: International Water Management Institute.

Kumar, M. Dinesh and O. P. Singh (2001) "Market Instruments for Demand Management in the Face of Scarcity and Overuse of Water in Gujarat," *Water Policy*, 3(5): 387–403.

———— (2005) "Virtual Water in Global Food and Water Policy Making: Is There a Need for Rethinking?" *Water Resources Management*, 19: 759–789.

Kumar, M. Dinesh, Katar Singh, and O. P. Singh (2001) "Groundwater degradation and its Socio-ecological Consequences in Sabarmati River Basin," Monograph No. 2, INREM Foundation, Anand, Gujarat.

Kumar, M. Dinesh, M. V. K. Sivamohan, and N. Bassi (2013) *Water Management, Food Security and Sustainable Agriculture in Developing Economies*, Routledge, UK: Earthscan.

Kumar, M. Dinesh, M. V. K. Sivamohan, and A. Narayanamoorthy (August 3, 2012) "The Food Security Challenge of the Food-Land-Water Nexus in India," *Food Security*, 4(4): 539–556.

Kumar, M. Dinesh and Jos van Dam (2009) "Improving Water Productivity in Agriculture in India: Beyond 'More Crop per Drop'," in *Water Productivity Improvements in Indian Agriculture: Potentials, Constraints and Prospects, Strategic Analysis of National River Linking Project of India Series 4*, edited by Kumar, M. Dinesh and Upali Amarasinghe, Colombo: International Water Management Institute.

Kutch Development Forum (1993) *A Demand for Review of Water Allocation to Kutch- Recommendations, Observations, Extracts from Report of Review Group on Narmada*, Bombay: Kutch Development Forum.

Malik, R. P. S. (December 27–30, 2006) "Do Dams Help Reduce Poverty?" Paper presented at the International Conference on Statistics and Informatics in Agricultural Research, New Delhi.

——— (April 2–4, 2008) "Growth Impacts of Development and Management of Water Resources," Proceedings of the 7th Annual Partners Meet, IWMI TATA Water Policy Research Program, ICRISAT, Patancheru, Hyderabad, India, pp. 858–869.

Matli, M. B. (December, 2005) "The Social Impacts of Large Development Project: Lesotho Highlands Water Project," Research report presented in Department of Geography, Faculty of Natural and Agricultural Sciences, University of the Free State, Bloemfontein.

Mc Cully, Patrick (1996) *Climate Change Dooms Dams, Silenced Rivers: The Ecology and Politics of Large Dam*, London: Zed Books.

Mehta, Lyla (1997) "Water, Difference and Power: Kutch and the Sardar Sarovar (Narmada) Project," IDS Working Paper No. 54, Institute of Development Studies, Brighton.

Mehta, Meera and Dinesh Mehta (2011) "Urban Drinking Water Security and Sustainability in Gujarat." In Volume 3 of *Sardar Sarovar Project on the River Narmada: Impacts so Far and Ways Forward*, edited by R. Parthasarathy and Ravindra H. Dholakia, pp. 727–743. New Delhi: Concept Publishing Company.

Mehta, Pooja (2005) "Internally-Displaced Persons and the Sardar Sarovar Project: A Case for Rehabilitative Reform in Rural Media," *American University International Law Review*, 203.

Meinzen-Dick, R. (1995) "Timeliness of Irrigation: Performance Indicators and Impact on Agricultural Production in the Sone Irrigation System, Bihar," *Irrigation and Drainage Systems*, 9: 371–387.

Merrey, D. J. (1996) "Institutional Design Principles for Accountability in Large Irrigation Systems," Research Report No. 8, International Water Management Institute, Colombo, Sri Lanka.

Modi, Narendra (2010) *Convenient Action: Gujarat's Response to the Challenges of Climate Change*, New Delhi: MacMillan Press.

Molle, François and Hugh Turral (June 2004) "Demand Management in a Basin Perspective: Is the Potential for Water Saving Over-estimated?" Paper prepared for the International Water Demand Management Conference, Dead Sea, Jordan.

Morse, Bradford and T. R. Berger (1992) *Independent Review of the Sardar Sarovar Project*, Ottawa, Canada: Resource Futures International.

Mukherjee, Sacchidananda, Zankhana Shah, and M. Dinesh Kumar (2010) "Sustaining Urban Water Supplies in India: Increasing Role of Large Reservoirs," *Water Resources Management*, 24(10): 2035–2055.

Narayanamoorthy, A. and R. S. Deshpande (August 30, 2003) "Irrigation Development and Agricultural Wages: Analysis Across States," *Economic and Political Weekly*, 38(26): 3716–3722.

National Rain-fed Area Authority (NRAA) (2011) *Challenges of Food Security and Its Management*, New Delhi: National Rain-fed Area Authority.

National Statistical Organization (NSO) (2012) *Energy Statistics 2012*, New Delhi: Central Statistical Office/National Statistical Organization/Ministry of Statistics and Programme Implementation/Government of India.

Pan, J. Z. and Z. Z. Zhang (eds) (2001) "Issues on Rational Allocation of Water Resources in Northern China and South–North Water Transfer. Strategies for Sustainable Development of Water Resources in China," Report No. 4, Water Publishing House, Beijing, China.

Paranjape, S. and K. J. Joy (1995) *Sustainable Technology Making Sardar Sarovar Project Viable: A Comprehensive Proposal to Modify the Project for Greater Equity and Ecological Sustainability*. Ahmedabad: Environment and Development Series, Centre for Environment Education.

Paranjpye, Vijay (1990) *High Dams on the Narmada: A Holistic Analysis of the River Valley Projects*, New Delhi: INTACH.

Parasuraman, S., Himanshu Upadhyaya, and Gomathy Balasubramanian (2010) "Sardar Sarovar Project: War of Attrition," *Economic and Political Weekly*, XLV(5): 39–48.

Patel, Anil (2001) "Resettlement in the Sardar Sarovar Project: A Cause Vitiated," *Water Resources Development*, 17(1), 315–328.

——— (2011) "What Do the Narmada Valley Tribals Want?" In Volume 2 of *Sardar Sarovar Project on the River Narmada History of Rehabilitation and Implementation*, edited by R. Parthasarathy and Ravindra Dholakia, New Delhi: Concept Publishing Company Pvt. Ltd.

Patel, Anil and Ambrish Mehta (2011) "The Independent Review: Was It a Search for the Truth?" In Volume 2 of *Sardar Sarovar Project on the River Narmada History of Rehabilitation and Implementation*, edited by R. Parthasarathy and Ravindra Dholakia, New Delhi: Concept Publishing Company Pvt. Ltd.

Perry, Chris J. (2001a) "World Commission on Dams: Implications for Food and Irrigation," *Irrigation and Drainage*, 50: 101–107.

——— (2001b) "Water at Any Price? Issues and Options in Charging for Irrigation Water," *Irrigation and Drainage*, 50(1): 1–7.

Priscoli, J. D. and J. Briscoe (2011) "Invited Opinion Interview: Two Decades at The Centre Of World Water Policy," *Water Policy*, 13(2): 147–160.

Raj, P. A. (1991) *Erroneous Zones, All About Narmada*, Gandhinagar: Government of Gujarat.

Rangachari, R. (2005) *The Bhakra-Nangal Project: Socio-Economic and Environmental Impacts*, New Delhi: Oxford University Press.

Rao, V. V. K. (September, 2006) "Hydropower in the North East: Potential and Harnessing Analysis," Background Paper No. 6, input to the study "Development and Growth in Northeast India: The Natural Resources, Water, and Environment Nexus."

Raskin, P., P. Gleick, P. Kirshen, G. Pontius, and K. Strzepek (1997) *Water Futures: Assessment of Long-Range Patterns and Problems*, Stockholm: Stockholm Environment Institute.

Robinson, Courtland (2003) *Risks and Rights: The Causes, Consequences, and Challenges of Development-Induced Displacement*, Washington, DC: Brookings Institution-SAIS Project on Internal Displacement.

Rosegrant, M.W., C. Ringler, and R. Gerpacio (1999) "Water and Land Resources and Global Supply". In *Food Security, Diversification and Resource Management: Refocusing the Role of Agriculture*, edited by G. H. Peters and J. von Braun. Proceedings of the 23rd International Conference of Agricultural Economics held at Sacramento, California, August 10–16, 1997. England: University of Oxford.

Rosegrant, Mark W. and Hans P. Binswanger (1994) "Markets in Tradable Water Rights: Potential for Efficiency Gains in Developing Country Water Resource Allocation," *World Development*, 22(11): 1613–1625.

Rosegrant, Mark W. and Renato Gazmuri Schleyer (1994) "Reforming Water Allocation Policy through Markets in Tradable Water Rights: Lessons from Chile, Mexico, and California," EPTD Discussion Paper No. 2, International Food Policy Research Institute, Washington, DC.

——— (1996) "Establishing Tradable Water Rights: Implementation of the Mexican Water Law," *Irrigation and Drainage Systems*, 10(3): 263–279.

Sadoff, Claudia and David Grey (April 18, 2005) "Water Resources, Growth and Development," Working Paper for Discussion, prepared by The World Bank for the Panel of Finance Ministers, the UN Commission for Sustainable Development.

Sah, D. C. (1993) Monitoring and Evaluation of Rehabilitation and Resettlement Programme for Sardar Sarvar (Narmada) Project, Report 16. Centre for Social Studies, Surat.

——— (1994). Monitoring and Evaluation of Rehabilitation and Resettlement Programme for Sardar Sarovar (Narmada) Project, Report 19. Centre for Social Studies, Surat.

Sah, D. C. (1996). Monitoring and Evaluation of Rehabilitation and Resettlement Programme for Sardar Sarvar (Narmada) Project, Report 21. Centre for Social Studies, Surat.

Sah, D. C. and Shubhra Tomar (2011) "Sardar Sarovar Project: Some Contentious Issues," in Volume 2 of *Sardar Sarovar Project on the River Narmada History of Rehabilitation and Implementation*, edited by R. Parthasarathy and Ravindra Dholakia, New Delhi: Concept Publishing Company Pvt. Ltd.

Scudder, Thayer (2005) *The Future of Large Dams: Dealing with Social, Environmental, Institutional and Political Costs*, London: Earthscan/Routledge.

Seckler, David (1996) "New Era of Water Resources Management: From Dry to Wet Water Savings," Research Report No. 1, International Irrigation Management Institute, Colombo, Sri Lanka.

Shah, Tushaar (1998) *The Deepening Divide: Diverse Responses to the Challenge of Groundwater Depletion in Gujarat*, Anand: The Policy School Project.

Shah, Tushaar, Ashok Gulati, P. Hemant, Ganga Shreedhar, and R. C. Jain (2009) "Secret of Gujarat's Agricultural Miracle After 2000," *Economic and Political Weekly*, XLIV(52): 45–55.

Shah, Tushaar and Barbara van Koppen (2006) "Is India Ripe for Integrated Water Resources Management? Fitting Water Policy to National Development Context, Special Articles," *Economic and Political Weekly*, August 5.

Shah, Zankhana and M. Dinesh Kumar (2008) "In the Midst of the Large Dam Controversy: Objectives, Criteria for Assessing Large Water Storages in the Developing World," *Water Resources Management*, 22(12): 1799–1824.

Siebert, S., J. Burke, J. M. Faures, K. Frenken, J. Hoogeveen, P. Doll, and F. T. Portmann (2010) "Groundwater Use for Irrigation—A Global Inventory," *Hydrology and Earth System Sciences*, 7(3): 3977–4021.

Singer, M. B. (2007) "The Influence of Major Dams on Hydrology Through the Drainage Network of the Sacramento River Basin, California," *River Research and Applications*, 23(1): 55–72.

Singh, O. P. (June 5–9, 2004) "Water-intensity of North Gujarat's Dairy Industry: Why Dairy Industry Should Take a Serious Look at Irrigation?" Paper presented at the Second International Conference of the Asia Pacific Association of Hydrology and Water Resources (APHW 2004), Singapore.

Skinner, J., M. Niasse, and L. Haas (eds) (2009) "Sharing the Benefits of Large Dams in West Africa," Natural Resource Issues No. 19, International Institute for Environment and Development, London, UK.

SSPL (2012). Sardar Sarovar Narmada Nigam Limited, Gandhinagar.

SSPY (2012). Here SSPY Stand for Sardar Sarovar Punaradhivas Yojana.

Talati, Jayesh and M. Dinesh Kumar (November 22–25, 2005) "Quenching the Thirst of Gujarat through Sardar Sarovar Project," Paper presented at the XII Word Water Congress, New Delhi, India.

Talati, Jayesh and Tushaar Shah (2004) "Institutional Vacuum in Sardar Sarovar Project: Framing Rules-of-the Game," *Economic and Political Weekly*, July 31.

Thakker, Mahesh (1988) *Sindhu Waters and Kutch*, Bhuj: Aaradhana Trust.

Tortajada, C. (2002) "Evaluation of Actual Impacts of the Atatürk Dam," *International Journal of Water Resources Development*, 16(4): 453–464.

Tortajada, C., H. D. Altibilek, and A. K. Biswas (2012) *Assessing Impacts of Large Dams*, Berlin: Springer Verlag.

Tortajada, C. and A. K. Biswas (2012) "Impacts of Large Dams", Global-Is-Asian, 14: 29–30.

United Nations Development Programme (UNDP) (2006) *Human development Report-2006*, New York: United Nations.

Vaidya, Prerna Desai (2011) "Irrigation Impacts of Sardar Sarovar Project: A Field Evaluation," in *Sardar Sarovar Project on the River Narmada Impacts So Far and the Ways Forward*, edited by R. Parthasarathy and R. Dholakia, Ahmedabad, Gujarat: Concept Publishing Company Pvt. Ltd.

van den Bergh, Jeroen (2007) "Energy Policy," Lecture Notes prepared for the Course in Environmental Management and Policy, Department of Economics and Business Administration, Free University, Amsterdam, the Netherlands.

van der Hoek, W., S. G. Feenstar, and F. Konradsen (2002) "Availability of Irrigation Water for Domestic Use in Pakistan: Its Impact on Prevalence of Diarrhea And Nutritional Status of Children," *Journal of Health, Population and Nutrition*, 20(1): 77–84.

Vaux Jr., H. J. and W. O. Pruitt (1983) "Crop-water Production Functions," in Volume 2 of *Advances in Irrigation*, edited by D. Hillel, Orlando, Florida: Academic Press.

Verghese, B. G. (2001) "Sardar Sarovar Project Revalidated by Supreme Court," *International Journal of Water Resources Development*, 17(1): 79–88.

Vyas, Jay Narayan (2001) "Water and Energy for Development in Gujarat with Special Focus on the Sardar Sarovar project," *International Journal of Water Resources Development*, 17(1): 37–54.

Watt, Jacqueline (March, 2008) "The Effect of Irrigation on Surface-ground Water Interactions: Quantifying Time Dependent Spatial Dynamics in Irrigation Systems," PhD thesis, School of Environmental Sciences, Faculty of Sciences, Charles Sturt University.

WELL (1998) *Guidance Manual on Water Supply and Sanitation Programmes*, Loughborough: WEDC.

White, G., D. Bradley, and A. White (1972) *Drawers of Water: Domestic Water Use in East Africa*, Chicago, IL: University of Chicago Press.

Whittington, D. and K. Chao (1992) "Economic Benefits Available From The Provision of Improved Potable Water Supplies: A Review and Assessment of the Existing Evidence." Wash Technical Report No. 77, Office of Health, Bureau for Research and Development, U.S. Agency for International Development under Wash Task No. 056.

WHO and UNICEF (2000) *Global Water Supply and Sanitation Assessment 2000 Report*, Geneva/New York: WHO/ UNICEF.

World Bank (2004) *World Development Report 2005: A Better Investment Climate for Everyone*, Washington, DC: World Bank/Oxford University Press.

———— (2006a) *Managing Water Resources to Maximize Sustainable Growth: A Country Water Resources Assistance Strategy for Ethiopia*, Washington, DC: World Bank.

———— (2006b) *Hazards of Nature, Risk to Development: An IEG Evaluation of World Bank Assistance to Natural Disasters*, Washington, DC: Independent Evaluation Group.

World Commission on Dams (2000) *Dams and Development: A New Framework for Decision Making*, London: Earthscan.

World Water Council (2000) "Ministerial Declaration of the Hague on Water Security in the 21st Century," the Hague, the Netherlands. Available at www.waternunc.com/ gb/secwwf12.htm (accessed April 3, 2010).

Yang, H. and A .J. B. Zehnder (2005) "The South North Water Transfer Project in China: An Analysis of Water Demand Uncertainty and Environmental Objective in Decision Making." *Water International*, 30(3): 339–349.

Index

Alagh, Y. K., 161
Allen, R. G., 258

Behar, S. C., 35
Berger, T. R., 33, 237
Bhakra Dam, Punjab, 24, 27
Buendia-Entrepenas reservoir, 17n1

CAD. *See* command area
    development (CAD)
canal irrigation, socioeconomic
    impact of
    agricultural scenario, 96–100
    area under irrigation, 100–102, 103f
    cropping intensity, 103–108, 104t,
        106t–107t
    cropping pattern, 103–108, 104t,
        106t–107t
    farmers, socioeconomic conditions
        of, 148–153, 150t–152t
    farming enterprise, 131–138,
        132f–134f, 135f, 136t–137t
    farm laborers, wage employment
        of, 147–148, 148t, 149f
    farm surplus, 139–146, 140t–146t
    gross cropped area, 103–108,
        104t, 106t–107t
    income from crop production,
        120–131, 121t–130t
    livestock holding of farmers,
        138–139
    sources of irrigation, 100–102,
        101f, 102f

water and agricultural scenario,
    96–100
water resources, 96
yield and returns of crops,
    108–120, 109t–110t, 110f,
    112t–115t, 116f–118f,
    119t–120t
carbon emission, 20
CF. *See* consumptive fraction (CF)
Chakravorty, Ujjayant, 30, 258
Chao, K., 212, 213
command area development (CAD),
    14
consumptive fraction (CF), 31
*Convenient Action: Gujarat's
    Response to the Impacts of
    Climate Change,* 31n4
crop yield, 281
cropping intensity, 103–108, 104t,
    106t–107t
Cully, Mc, 17

Daines, S. R., 30
Dharoi reservoir, 32n5
Dinesh, Kumar, M., 30
displaced population
    drinking water sources, 249–250,
        250t
    farming enterprise, 245–248
    farm land, ownership of, 244–245
    health facilities, 250–251, 251t
    new settlement, 254, 254t

rehabilitation status, 240–244, 243t
relocation, socioeconomic impact of, 244–254
R&R policy, 236–240
stress, 251–254
technology adoption, 249
drinking water
fluoride and salinity, 2
drinking water supplies from Narmada
current scenario, 160–161
domestic water consumption, 169–171, 171t, 172t–174t
drinking water quality, 175–177, 175t–177t
and irrigation, 183–184
socioeconomic impact, 161–164, 161t, 163t
water collection, time spent on, 168–169, 169t
water-related diseases, 177–179, 178t–179t
water supply, frequency of, 164–166, 165t
water supply, source of, 166–168, 166t, 167t
Duflo, Ester, 23

energy demand, 4
environmental externalities
carbon emission, reduction in, 224
electricity, sale of, 223f, 224
fertilizer and pesticide application, 227–228
Narmada water, ecological benefits of, 228–234
salinity of groundwater and soils, 224–225
waterlogging, 226–227
environmental stresses, 3
ET. See evapotranspiration (ET)
evapotranspiration (ET), 29–30

Fisher, William. F., 235, 236
Frederick, K. D., 268

Gleick, Peter H., 2
global context, large water projects
environmental impacts of, 17–29
externalities of, 29–38
socioeconomic impacts of, 17–29
GRA. See Grievance Redressal Authority (GRA)
Granit, Jacob, 18
Grey, D., 5, 157
Grievance Redressal Authority (GRA), 34
groundwater
geo-hydrological impacts, 78–83
in Gujarat, 76–78
hydrogeology of, 83–87
pre-Narmada vs. post-Narmada, 90–94
quality, 2
surface water, 87–89
surface water and, 4
Gujarat
agricultural gross domestic product (GDP), 8
agricultural growth in, 1
drinking water, acute shortage of, 2
economic and socioeconomic impacts, 7
in groundwater, 76–78
groundwater levels, 1

ICESCR. See The International Covenant on Economic Social and Cultural Rights (ICESCR)
ILO. See International Labor Organization (ILO)
institutional approach
fixing volumetric water entitlements, 268–269
irrigation water pricing, 267–268
water trading, 268–269

institution building, 269–271
Integrated Water Resources
    Management (IWRM), 22
International Covenant on Economic
    Social and Cultural Rights
    (ICESCR), 34
International Labor Organization
    (ILO), 34
Irajpoor, A. A., 26
irrigation and drinking water supply
    canal, impact of, 212
    dairy production, 197–204, 200t
    domestic water supplies, 207–211,
        210t, 211f
    drinking water provision, social
        benefits of, 212–213, 213t
    electricity for, 214–216, 215t
    energy saving benefits, 185–187
    groundwater availability, 185, 186f
    wage increase, 204–207, 205t, 206f
    water level fluctuations, 185, 186f
    well-deepening, cost of, 188, 188t
    well irrigation, 189, 190t, 191,
        192t, 193t, 194, 194f, 195t–196t
    well irrigation, canal irrigation,
        202–204
IWRM. See Integrated Water
    Resources Management (IWRM)

Jain, L. C., 35

Kumar, M. Dinesh, 1, 5, 8, 19, 77, 98,
    157, 171, 220

LAA. See Land Acquisition Act
    (LAA)
labor-intensive irrigation techniques,
    25, 26
Land Acquisition Act (LAA), 34
large dams, 19
Latif, M., 26
Lesotho Highlands Water Project
    (LHWP), 20, 21

LHWP. See Lesotho Highlands Water
    Project (LHWP)
life-sustaining value, 3
Lindstrom, Andreas, 18

Malik, R. P. S., 23, 27
Matli, M. B., 20
Mehta, Ambrish, 34, 237
micro-irrigation (MI) systems, 8,
    265–266
Morse, Bradford, 33, 237
Mukherjee, Sacchidananda, 28

Narmada Main Canal (NMC), 64
Narmada river basin
    demand-driven water scarcity, 67
    drinking water supplies from,
        159–160
    environmental water scarcity, 68
    hydropower generation, 71
    physical scarcity of water, 65–66
    progress in implementation, 71–73
    Sardar Sarovar project on, 64–65,
        68–73
    water and power systems for,
        62–64
    water distribution system of SSP,
        70–71
Narmada Water Disputes Tribunal
    Award (NWDT Award), 34, 35
NMC. See Narmada Main Canal
    (NMC)
NWDT Award. See Narmada Water
    Disputes Tribunal Award (NWDT
    Award)

PAFs. See project-affected families
    (PAFs)
Pande, Rohini, 23
Pan, J. Z., 229
Participatory rural appraisal (PRA)
    tools, 40
Patel, Anil, 34, 237, 238, 240

Pawar, J. R., 30
politics of dams and development, 16
politics of hydrology, 16–17
PRA tools. *See* Participatory rural
  appraisal (PRA) tools
project-affected families (PAFs), 13,
  14
Punjab, Bhakra Dam, 24, 27

Ranade, Rahul, 77
Raskin, P., 68
reservoir storage, development of,
  291
Resettlement and rehabilitation
  (R&R) policy, 13
Robinson, Courtland, 35
R&R policy. *See* Resettlement and
  rehabilitation (R&R) policy

Sadoff, C., 5
Sah, D. C., 244, 249, 252
SAM. *See* Social Accounting
  Matrices (SAM)
Scudder, Thayer, 236
Shah, Zankhana, 1, 19, 220
Singer, M. B., 26
Skinner, J., M., 21
Social Accounting Matrices (SAM),
  24
social benefits and impacts, SSP
  approach, 40–46
  canal irrigation, positive
    externality of, 49
  clean energy production, 51–52
  crop and dairy production, 46–48
  data, type and sources of, 52–54,
    57–59
  direct and indirect benefits from,
    276–277
  drinking water supply benefits, 50

energy saving benefits, 50
  farm surplus, 48–49
  sampling procedure, 52–54
  study locations, 54–55
  surface water for drinking, 51
static water levels (SWL), 53
Sustainable Water Use Index (SWUI),
  19, 291–292
SWL. *See* static water levels (SWL)
SWUI. *See* Sustainable Water Use
  Index (SWUI)

Talati, Jayesh, 171, 216
TDS, 42
Tomar, Shubhra, 244, 252
Tortajada, C., 36

Umetsu, Chieko, 30

van Dam, Jos, 30
village service area (VSA), 71n3

water-intensive crops, 23
water projects, direct and indirect
  benefits of, 6–15
water-related disasters, 274
water-rich areas, of South and Central
  Gujarat, 264–265
water-scarce areas, of North Gujarat,
  262–264
Water security and economic growth,
  3–6
watershed management, 5
water storage technology, 18
water-stressed regions, 3
water users associations (WUAs), 14
Whittington, D., 212, 213
World Bank, 17

Zhang, Z. Z., 229

About the Authors

S. Jagadeesan retired from the Indian Administrative Service in 2013 as the Managing Director of Sardar Sarovar Narmada Nigam Limited (SSNNL). He has held many prestigious positions in state and central government services as a bureaucrat, and has also served as Minister (Economic) to High Commission of India, London.

M. Dinesh Kumar is Executive Director, Institute for Resource Analysis and Policy (IRAP), Hyderabad. He has 24 years of professional experience in the water and agriculture sector, undertaking research, consulting, action research, and training. He is also the Associate Editor for *Water Policy* and the Editorial Board Member for *International Journal of Water Resources Development*. He has three books and nearly 150 research articles, including many in international peer-reviewed journals, to his credit.